DECODING THE HUMAN MESSAGE

HENRI LABORIT

Decoding the Human Message

translated by
Stephen Bodington & Alison Wilson

Allison & Busby, London

First published in Great Britain 1977
by Allison and Busby Limited
6a Noel Street, London W1V 3RB

Copyright 1. © Editions Robert Laffont 1974
Translation copyright © Allison & Busby 1977
(Original title: *La Nouvelle Grille*)

ISBN 0 85031 186 1 (hardback)
ISBN 0 85031 187 X (paperback)

Set in Times 10 on 11pt and printed by
Villiers Publications Ltd, Ingestre Road, London NW5

Contents

Introduction

Man has never been able to do without a decoding "grid" of some kind. Faced with the apparent disorder of the world, he has had to look for the meaningful terms which, in association with each other, would make his action on his environment more effective and enable him to survive. Faced with the infinite abundance of objects and beings, he has searched for the relationships between them; and faced with the infinite mobility of things, he has sought invariance.

He has also added information to the inanimate world. This information could only consist of what already structured his nervous system. Fortunately, the combinatorial pool granted him by his associative systems opened up the path of language, abstraction and the creation of new frameworks for interpreting the world: in other words, the path of complexity. In fact, once his interpretative frameworks had been rewarded with the increased effectiveness of his action on the environment, a new store of facts which had lain unknown until then began to complement the harvest of knowledge accumulated over previous centuries. Man has always in some way or other proceeded by means of working hypotheses followed by experimentation. But for millenia, what he could not explain by logical reasoning remained in the domain of myth. It made the priesthood their fortune.

The discovery of the steam engine developed man's interpretative pretentions. And it was a fact that the discovery of the laws of the inanimate world accelerated to such an extent, from a certain point onwards, that in a few decades man had become the master of energy. But his knowledge of himself did not accelerate to the same extent, and today he disposes of considerable powers of destruction in complete ignorance of his own unconscious. He must face up to the evidence: the thermodynamic grid by itself does not enable him to explain the living world.

An integral part of this world of the living is made up of the individual, the species, and groups of individuals — the many social sub-sets which are scattered over the world.

For centuries people have tried to decode the complex and

7

dynamic structure of social relations. The most recent grids, those of Marx and Freud, have afforded a more precise and therefore a more effective reading of this structure. Unfortunately, they still prevent us from getting to know the complex mechanisms which govern the functioning of the human organism, and of the nervous system in particular. And it is these which act as the intermediary for inter-individual relations.

This gap is incomprehensible. The living systems are constituted of the same atomic elements as inanimate matter: it is their structure which is different. A scientific approach to these structures cannot do without the notion of information and of cybernetic regulation. And these notions have been in existence for scarcely twenty years. We learned to use them effectively in biology at an even later date.

The "new grid" is therefore the biological grid which gives us an insight into the complexity of our behaviour in the social situation. I have come to learn about this gradually in the course of my day-to-day experience in the laboratory, and I have tried to explore the possibilities for interpreting social phenomena which this grid seems capable of providing us with. Most of the books which I have written over the past twenty years spring from my experimental work on the "organic reaction to aggression". This work has led me to make certain conceptual syntheses, and these appear in my latest specialised work.[1] This tries to gather together the essentials of what we know about the functioning of the human nervous system, and about the enzymatic reaction to behaviour in the social situation.

My approach has thus above all been experimental.

But the social importance of this knowledge seemed to me to be so great that it had to be prevented from being imprisoned behind laboratory walls. So at the same time I have written popularised works in the hope of disseminating certain ideas which seem to me to be indispensable for every one of my contemporaries, and which may enable them to situate themselves better in the world in which they live, in relation to other people and in relation to themselves. Perhaps I have also been hoping to take part in the development of contemporary society.

What I propound in this present work is a new grid* for decod-

---

*There is a rich ambiguity in this term — more evident in the French *grille* — which is central to the author's theme. It must be understood *both* as a "decoder" (cf. the English "reference grid") permitting interpretation *and* as a barrier, something which encloses [*trans.*].

ing human experience, although I know that it will only be a "new" one until another more complete and comprehensive grid replaces it, as this one seeks to comprehend the preceding ones.

This biological grid is also, in my view, as generalising and as interdisciplinary as possible in the domain of biology itself. The reader will recognise in passing many borrowings which I have made from ethology, from behaviourism or the Skinnerian study of behaviour, from psychoanalysis, experimental psychology or contemporary linguistics. He will find certain ideas which I have developed in other books over the past ten years, or in experimental work in our research group's laboratory. But the assimilation of a scientific fact in a set which includes it usually transforms it profoundly. We encounter more than a simple interdisciplinary synthesis: it is the interdisciplinary construction of a new set, which transfigures each element by creating new relationships in it. Thus aggression in man, as it is presented to us by certain ethologists such as Konrad Lorenz and Robert Ardrey, changes its meaning completely when it is placed in a broader biophysiological framework. The same is true of affectivity: since McLean it has been supported by the limbic system, but in the broader framework it becomes a by-product of the memory. The memory in its biochemical form becomes the basis for sub-cultural automatisms, and the latter become the turntable for conflicts between the hypothalamic drives and the building of imaginary structures.

It was Montaigne who suggested that we should follow the bees' example, by gathering the pollen from a variety of flowers precisely in order to do something different with it. I hope our honey is edible. It must be sampled at the sociological, economic and political levels. In fact the ideas of information, of dynamic structures and of the open system at the thermodynamic and informational level may profoundly alter the automatisms of thought which we usually entertain on this topic.[2] Many of the problems which I deal with in this book, and which are a resumption of previous writings, have recently become fashionable subjects which are dealt with in the mass media and are tossed about at cocktail parties. But they are usually treated in a pigeon-holed form, because it is only the specialist who has any impact on opinion: he has to be believed, because his explanation necessarily appears in a simplified form — that of his own discipline. Syntheses are usually more complex. They demand of the reader a great deal of attention and a glimpse of the knowledge which stretches beyond his horizon, even when they are popularised or simplified. It is therefore more difficult to disseminate them. Their conclusions too are profoundly different.

And above all, when a synthesis cannot be easily integrated in known cultural schemas or when it does not favour a pre-existing current of opinion, a currently fashionable political or socio-economic ideology, it has little chance of finding an immediate echo. It cannot be taken seriously. The person who proposes it cannot be trusted: he is not a fine-grained humanist, with the kind of humanism which lets things alone and refers to the great heritage of the past, the existing culture and the entire set of prejudices that link a society with an epoch.

But the establishment of a new structure is a cause of such personal joy that approval is neither a necessary nor a sufficient motivation, even if it is pleasant.

It may be difficult to summarise the book, but I would like to try, so that the reader is not disconcerted by the first two chapters. If his culture is mainly a literary one, he will be in danger of closing the book and leaving it aside. Take the trouble to read the first two chapters, even if the terminology sometimes appears difficult. They are basic to the whole, since they try to present a dynamic vision of the organisation of living systems on the basis of recent information theory, the theory of systems and cybernetics. They will subsequently be clarified by the chapters which follow, beginning with the third chapter, where the argument is presented at the sociological level to which this book as a whole is directed. The second chapter outlines the functional structure of the nervous system. One cannot, in fact, ignore this marvellous instrument: we use it from the moment we are born until we die, and it sanctions all our relationships with the world around us, the world which is peopled by other human beings and which gives us a consciousness of the world within us. But this world within, the world of the unconscious, is quite different from the consciousness we have of it: it is structured chiefly by molecules, not simply by words.

These are the conceptual tools with which I deal with the theme as a whole, in its social, economic and political aspects. To my knowledge they have never been employed before in the experimental approach to social phenomena; they have emerged only recently, and the synthesis which I make of them is something difficult for the specialist to achieve.

I have often been accused of seeking analogies with the biological in the social. To reason by analogy is rightly something which has a bad reputation. But perhaps those who make the accusation are simply trying to shake off certain annoying facts, and to remain within the restricted territory which is theirs and which gives them a certain gratification. I sometimes use analogy, but the essentials of what I have to contribute do not in my understanding belong to the realm of analogy. In fact, the observation of biological facts has in my opinion led us to a discovery of structural laws which are valid for the realm of the living as a whole. To reproach someone for making what appears to be an analogy, according to a superficial examination or emotional judgement, is like reproaching someone for the fact that the laws of gravity apply to a parachutist. I will quite happily agree that traffic does not circulate in a town because there are "arteries", nor because one can compare a town to an organism or the "heart of a city" to the heart of a mammal: that is a job for the poets (and for some town planners).

Giving a precise definition to the notions of energy, mass and information enables us to base sociology, economics and politics on a firm tripod, because the former support the edifice of contemporary science. But in order that this tripod may reach the level of the so-called "human" sciences, it must also serve as the basis for locating general biology and the biology of human behaviour in the social situation. We shall therefore encounter biology on every single page of the eleven chapters which follow the description of the nervous system.

We shall see how the fact that man possesses an orbito-frontal lobe and developed associative systems in the cortex enables him to process information, and by what mechanisms his imagination adds information to the world around him. We shall see how this specific attribute was at the origin of his domination of the inanimate world, and later became the basis of hierarchies of domination which were based solely on the individual's degree of technical and occupational information. We shall see why animal societies and human societies are subject to this pressure of necessity which the hierarchical structures constitute. We shall analyse the mechanisms by which power and domination are established, the notion of territory and property, and the myths of democracy, equality and freedom — words which simply express an effective drive that is either gratified or, on the contrary, alienated and subordinated to domination by the other.

I shall try to supply the beginnings of a solution. In order to do this, I shall explain the distinction between occupational information, which introduces the individual into the process of commodity production, and generalised information (to which this book

11

is a guide). Generalised information alone can give the citizen the status of human being. It concerns structures rather than facts. It also concerns the general laws which enable man to organise facts free from value judgements, socio-cultural automatisms, prejudices, morals, ethics (which can be defined as the strongest ethics which the police are capable of enforcing), war, legislation, the degrading effects of the mass media, economic alienation, emotional obscurantism, the blinding effects of the logic of speech, and above all the gratification provided by the occupational hierarchy. Generalised information forms the human nervous system into an open system, and it cannot be satisfied simply by the worn-out slogans and sham revolutionary phraseology which indicate a resumption of hierarchical domination and the gratification of power after the "destruction" of the capitalist structures.

To be a human being consists chiefly in using the cerebral regions which distinguish us from other animal species and enable us to create new structures. To be a revolutionary does not mean to apply decoding grids which were invented in a period when two thirds of our contemporary scientific knowledge had still to be discovered, a period moreover which remained confined within the interpretations of conscious language and logical analysis: these utilise the principle of linear causality and do not call in question the drives and automatisms which guided and still guide our discourse. To be a revolutionary means first of all to conceive new conceptual grids and new structures that can take over the essential contribution which the biological disciplines as a whole have made — not in detached bits, in the form of cultural bric-à-brac, but in an integrated form which starts out with physics and ends up with the human species in the biosphere. Time is the time of evolution and of the individual; space is the gratifying space of the human being and of all human beings — the planet.

To be a revolutionary is therefore no longer the affair of a handful of inspired leaders, an élite which enlightens the mass: it is the affair of all. To be a revolutionary is the finality of the human species: the revolution is both permanent and cultural, and this implies more than simply a social praxis or a linguistic culture. There can be no fruitful experimentation without a working hypothesis; but on the other hand no experiment can be restricted to verifying a theory which has not had the chance to take into account the fundamental laws discovered after it was proposed. Any theory which cannot adequately explain certain facts of experience must be included in a broader theory, where it will become a sub-

12

set: that is, of course, on condition that it gives a logical and above all a verifiable interpretation to many of the other sub-sets. If it does not do so, it is a myth and should be abandoned.

We shall see how dangerous words are, because the object or concept which they are supposed to represent is very quickly forgotten, and because they make us content with an unsatisfied affectivity and with exploiting the frustration which springs from our growing inability to carry out acts of gratification. Political thought seems to be increasingly encumbered by this kind of language.

Consciousness, knowledge and imagination are the only specific characteristics of the human species. They are also the characteristics which are least often used. On the contrary, man entertains a false idea of his species, and behind the profitable veneer of fine sentiments and great ideas, he ferociously maintains all the forms of domination. The only way to uproot these unfrocked ideas is to dismantle their mechanisms and to generalise knowledge.

Truth is a naked woman who emerges from a well: the well is the obscure fluency of our unconscious. This book is dedicated, without pity, to those who suffer: the poor, the alienated, the imprisoned, the addicted, the rebels, and all those who do not feel happy as they are. But it is also dedicated to well-off people, respectable people, cops, presidential candidates, distinguished citizens, and all those who feel sure they know the correct answer (whether it is of the right or of the left) — in the hope that they will discover here at least the germs of uncertainty, which is the sister of anxiety and the mother of creativity.

1

Thermodynamics and information: physics and biology

Until not so long ago, one merely had to state an opinion and one could rapidly be classified as either a materialist or a spiritualist. The materialists, reaping a centuries-old inheritance of experiments which have made man the privileged observer of the material world, thought they could interpret the organisation of living matter through the laws of physics alone. The spiritualists, enclosed in conscious discourse, declared that it was impossible to reduce what they called "life" to matter, and they endowed life with a secret energy, a "life-force" whose most refined expression was "the mind". Until laws were discovered for structures and sets, with the advent of information theory and cybernetics, it was impossible to understand that, in Norbert Wiener's words, living systems add neither mass nor energy to inanimate matter but simply information. While it is quite true that information must be supported by mass and energy, it is also true that information represents the "something" which makes the whole more than the sum of its parts. Living matter is made up of the same atomic elements as inanimate matter, but is distinguished from it by the particular organisation these elements acquire within it, and by the relationships operating among them. If structure is defined as the whole set [*ensemble*] of relationships existing among the elements of a set (see figure 1), then it is indeed the structure of living matter which is responsible for all its original characteristics. (Since the whole set of relationships among the elements of a set is beyond the reach of our consciousness, the word "structure" will be used to designate the sub-sets of the whole. Clearly, this structure depends on the observer making the abstraction.) Since structure cannot be weighed in a balance nor measured by a dynamometer, it has indeed the "immaterial" qualities of "the mind". But it cannot be separated from matter, any more than the signifier can be separated from the signified.

15

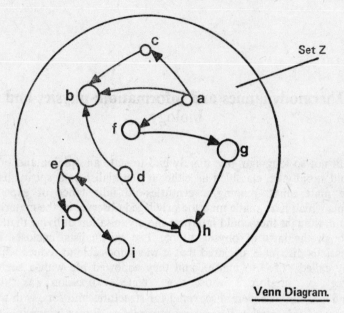

Venn Diagram.

a, b, c, d, e, f, g, h, i, j = elements of set Z

The arrows represent the relationships among the elements of this set.

Figure 1

Open systems and levels of organisation

The structure of living matter confers two basic characteristics on it: first, it is an open system; secondly, it is organised according to levels of complexity. Furthermore, these two characteristics are strictly dependent on each other.

Biological open systems

By talking of open systems, we admit that closed systems exist. There are two ways of considering whether a system is open or closed: one is at the thermodynamic level, the other at the informational level. Biological or physical regulation is often talked about, and a regulated system may sometimes be a closed system. But I would like to show that biological regulation is of a particular sort, and that it occurs in systems which are open from both the thermodynamic and informational points of view.

16

(a) The thermodynamic opening:

An enzymatic reaction converts one molecule (the substrate) into another molecule (the product of the reaction). This conversion might be possible without a third molecule, the protein molecule called an "enzyme", but in that case considerable energy would be needed to excite or agitate the molecules concerned. The conversion would depend on the chance meetings of the molecules, increased by the thermal agitation which is itself produced by the conversion of thermal energy into non-recoverable kinetic energy. The enzyme, an intermediary molecule whose form in space generally allows it to bond with one single substrate in an "active site", permits conversion of the substrate with the slightest possible expenditure of energy at temperatures accessible to biological processes.

An enzymatic reaction, isolated *in vitro,* is a system which, once it has achieved equilibrium, is regulated by the law of action of mass. This equilibrium depends on the quantity of substrates, the quantity of enzymes, and the quantity of the product of the reaction concerned. The reaction is written by using the sign $\underset{\longleftarrow}{\overset{E}{\longrightarrow}}$ placed between the substrate and the product of the reaction, above which E represents the enzyme which ensures the catalysis. Thus most enzymatic reactions are so-called reversible reactions. In cybernetic language it may be written:

$$\text{substrate (factor)} + E \xrightarrow{\qquad\qquad} \underset{\text{(effect)}}{\text{Product of the reaction}}$$

negative feedback

This indicates that the quantity of substance produced by the enzymatic reaction will negatively influence the quantity of substrate used. If no extra substrate is added and if all the physico-chemical conditions remain constant (pH, temperature etc.), this system remains in equilibrium. It is regulated to stay constant. Nothing else happens.

A living cell, however, is a chemical factory, where on the contrary a lot of things happen. This is because enzymatic reactions in a living cell are not isolated from one another, but succeed each other in a chain. The substrate of one reaction consists of the product of the previous reaction. Thus a substrate such as glucose goes in at the beginning of the chain and at the end a waste product emerges, carbon dioxide and water ($CO_2 + H_2O$), as a result of its degradation. But meanwhile the excitation energy of the electrons linked to the hydrogen (H_2) molecules contained in the substrate will have been gathered and put in reserve in the form of energy-rich phosphorus compounds (ATP, GTP, UTP, phosphocreatine), by which this energy ensures the labile linking of phosphorus atoms. This lability will permit rapid use of this potential energy for any cellular work (mechanical, electrical, secretory, or synthesising). A living system of any sort — a cell, an organ or an organism — is therefore, *from the thermodynamic point of view, an open system.* A current of chemical energy passes through the system from the original substrate to the waste products. This energy current is very often an electronic one, for the role of the enzyme is frequently to facilitate the exchange, one electron at a time, from the substrate to the product of the reaction. In many cases the oxido-reduction process is involved. The substrate is oxidised (it loses an electron) while the product of the reaction is reduced (it gains an electron). This product will lose the electron again, giving it to another molecule in the following enzymatic reaction. In the course of being passed around, the electron "cascades" from a high energy level, from an outer electron orbit, to an orbit closer to the nucleus, before returning to its base orbit. With each quantic leap it releases a small amount of energy. This will be held in reserve in the third phosphorus link of adenosine triphosphate (ATP).

Furthermore, the set of life-forms in the biosphere form a vast open system in which *solar energy* flows. Indeed the substrates I have just mentioned (the chemical molecules from which living organisms draw their energy "in small change", as Szent-Györgyi put it, describing the biological oxidations involved in the electrons' energy cascades, in contrast to the large "bank-notes" of non-biological combustions) are already complex molecules, constructed from solar photon energy with the help of the chlorophyll molecule. This molecule enables the sun's photon energy to be converted into chemical energy. Thus life-forms do not contradict the second law of thermodynamics, the Carnot-Clausius law, for it is because of solar entropy that living structures and the total amount of energy they release can be maintained (see figure 2). This global thermodynamic aspect can also be found in the human economy.

(b) The informational opening:
With a living "set" of any kind, from bacteria to human societies, if we talk of "living structures" we are talking about the set of relationships among the elements which make up this set. Therefore,

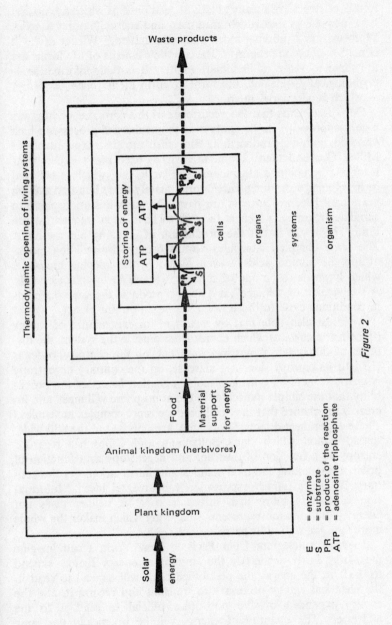

Waste products

Thermodynamic opening of living systems

Storing of energy

ATP   ATP

cells

organs

systems

organism

Food

Material support for energy

Animal kingdom (herbivores)

Plant kingdom

Solar energy

E = enzyme
S = substrate
PR = product of the reaction
ATP = adenosine triphosphate

Figure 2

19

to talk of structures means to talk of relationships which are neither mass nor energy, but which need mass and energy in order to exist. "Information is nothing more than information," Wiener said, "it is neither mass nor energy." The atomic elements of life-forms are the same as those of inanimate matter. It is their information — etymologically speaking, the form given to them [*mise en forme*] — which is peculiar to them.

One often reads that the occurrence of this form, i.e. of life, was highly improbable in the context of the haphazard processes which are said to have produced it. But since the first experiments of Miller, Oro and then Calvin, there have been many experiments using a reconstituted atmosphere similar to that which is believed to have existed on our planet approximately three hundred million years ago. Electric currents are passed through this atmosphere to simulate lightning, which is thought to have been frequent at the time. The result is not the production of just any old molecule. The same complex molecules always occur: the building blocks of life, i.e. amino acids — even including the adenine molecule, which happens to be one of the bases found in the nucleic acids. A strange sort of "chance", if it always produces the same materials in conditions close to those when life began on our planet.

We must also note that by means of an expenditure of energy (the electric current) which comes from outside the system, the disorder of the system is not increased but, on the contrary, order is created; its entropy does not increase, on the contrary negentropy is created. Increasing the molecular agitation increases the probability that the simple molecules in the atmosphere will meet, and increases the chance that they will produce more complex molecules.[3] These experimental facts, let us note, are opposed to the still widespread notion which since Boltzmann and Gibbs has regarded entropy as a function of disorder and negentropy as a function of order, thus uniting thermodynamics and information. On the contrary, Wiener's phrase expresses a fundamental idea: "Information is nothing more than information. It is neither mass nor energy." Information represents something which makes the whole more than the sum of its parts.

To send a telegram from Paris to New York, I can imagine measuring fairly accurately the amount of energy I must expend to draft it, the energy the post-office clerk will expend to send it, the electrical energy necessary to transmit and receive it, and the energy expended by the post-office official to send it to the addressee. The quantity of energy will be practically the same whether the letters making up the text are arranged in certain order which makes it meaningful to the person receiving it, or in disorder so that the telegram no longer carries information. Here we find the problem of Maxwell's demon again: *information needs mass and energy as its support, but it cannot be reduced to these two elements.* The signified is linked with the signifier, but one cannot be reduced to the other.

The most curious thing about the physico-chemical processes which brought about the first living structures is that they certainly seem to have obeyed laws, since we are already empirically capable of reproducing experimentally the first and already complex constituent parts of their specific order. This information, this giving of form, does not appear to be the result of chance alone. However, it should be noted that very specific conditions are needed to create order out of disorder. It cannot be achieved on an exploding star like our sun, nor on a heavenly body where molecular activity is almost non-existent because the temperature is near to absolute zero. The thermal range allowing the birth of the first life-forms is in fact very narrow. What are the physico-chemical laws which govern it?

As soon as the information leading to the first complex structures could be stored in the double helix or deoxyribonucleic acids (DNA), it permitted the reproduction of the same forms in many copies. In recent years molecular biology has provided us with basic knowledge about the process of transmission of information. This is not the place to consider this at length, but abundant cybernetic regulations are to be found in it. On the other hand, in order to understand what an open system is from the informational point of view, it may be interesting to consider why living organisms, whether plant or animal, are characterised by *cellular* organisation. Any living structure is a complex structure, built out of elements taken from the inanimate environment around it. Exchanges take place through the surface, and since the structure increases in volume cubically whereas its surface increases only as the square, the exchanges diminish as the mass increases, and then division becomes necessary. Here again, constant negative feedback regulates the system. This is what distinguishes living matter from inanimate matter: a crystal, for example, increases in surface and volume but not in complexity. What we call complexity in a living organism expresses the existence of different levels of organisation leading to the autonomy of the set within its surroundings. *It is these levels of organisation which permit the opening of the system at the informational level.*

Levels of organisation

Let us first choose an analogy from the world of inanimate matter, to describe this opening in the informational sphere. A water-bath with its thermostat regulated to maintain the temperature of the bath at 37°C, for example, is a *closed system*. It needs electrical energy so that the heating element heats the water, but if this energy remains constant then, according to the structural characteristics of the particular apparatus, as soon as the temperature departs from the average to any extent the heating circuit will open or close and re-establish the average temperature. From then on, except for oscillations of varying extent around the average temperature, due to time-lag and the hysteresis peculiar to the apparatus, nothing else will change. Feedback closes the system on itself.

This system is open at the thermodynamic level, since electrical energy is broken down into heat. But it is closed at the informational level.

A water-bath in a laboratory is a system which usually forms part of a more complex series of instruments set up for a complete experiment. The water-bath must often be regulated for temperatures other than 37°C. In this case information coming from outside the regulated system, from the experimenter, will change the level of regulation and establish it at 30°, 20° or 15°C for example. The finality of the closed system is then introduced into a more complex system, because of the information reaching it from this more complex set. We shall define the regulated system which receives information from outside itself, and changes its level of regulation, as a servomechanism (see figure 3) in the conventional way.

We have seen that as soon as the enzymatic reaction is isolated *in vitro* we have, in biological terms, a regulated system. Each of these systems is transformed into a servomechanism by the information received from the set which encompasses it. In this way the molecular group forming an enzymatic system receives its information from the chain of reactions of which it is part. This chain itself is situated in infracellular morphological structures (mitochondria, nucleus, endoplasmic reticulum, microsomes, etc.), whose activities depend on the information reaching the cellular

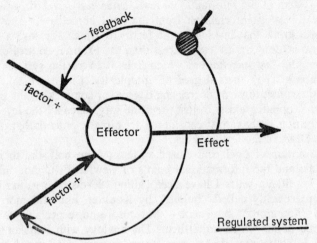

Figure 3

system through the nerves, the endocrine system, the ionic system, etc. The cell, which is demonstrated by the maintenance of the membrane's potential of rest to be a regulated system, becomes a servomechanism by virtue of the information reaching it from outside, which influences its functional activity by varying this potential. We can continue this description and find a regulated system and servomechanism at each level, from the cell to the group of cells forming an organ, and from the organ to the system it is part of (the nervous system, the cardiac, vascular or endocrine systems, etc.). Thus with each level of organisation up to the organic set, we can envisage it in an ideal state, i.e. the physiological state in which the information reaching it from outside (that is, from its environment) remains within narrow limits and in an intermediary state, where there is an organic reaction allowing flight or struggle, and where homeostasis is not preserved but motor autonomy in relation to the environment is. When this reaction of flight or struggle cannot act to re-establish the conditions favourable to life, we encounter what is known as a "pathological" state.

23

Whatever level of organisation we consider, from the molecule to the entire organism, we now know that an organic set is an open system at the informational level, since each level of organisation receives its information from the level immediately above it. We also know that in biology the feedback which "closes" a level of organisation carries less interest than the external control of the system, the *servomechanism* which links it to all the systems that encompass it by their degree of complexity. But if we take the opposite route, towards decreasing degrees of complexity, the informational opening exists in just the same way, since all the levels of organisation we have just considered are already potentially present in the DNA of the fertilised egg.

The notion of open and closed systems, from both the thermodynamic and the informational points of view, leads us not only to what for fifteen years I have been calling "levels of organisation" — more recently called "holons" by Koestler, and still more recently "integrons" by Jacob — but also and especially to the dynamic bonds which unite them. The analogy with Russian dolls which is sometimes drawn seems very imperfect to me, since it does not take the servomechanism into account. Indeed, even if one removes all the dolls from the biggest one, the latter would still keep its shape. The same action on a living organism would transform it into a corpse.

This approach leads us to the idea, essential in its sociological implications, that an organism is on the contrary a closed system as far as its "information-structure" is concerned.

Information-structure and circulating information

At this point I shall simply mention this vitally important distinction, for we shall return to it at greater length later on. The informational opening to which I have just referred, and which results from the structure in levels of organisation of living organisms, empowers what may be called "circulating information". This is carried at the cellular stage principally by "chemical messengers", the hormones, and by the nervous system. It is related to "information" as understood by telecommunications engineers. As in telecommunications the biologist must avoid the interference or "noise" which will distort the message. Moreover, this information requires a more or less specific reception structure in order to be decoded, for without this it would be meaningless.

But when I spoke of the information which "gives form" to an

24

organism and distinguishes it from the inanimate world, I did not mean this kind of information. We shall call the former "information-structure": this is what enables us to distinguish a man from an elephant, and this too must be protected from interference. But it does not circulate and it does not vary, at least as far as the individual is concerned. It is transmitted on another time-scale, through reproduction and the genetic code. From the point of view of his information-structure, the individual may be considered more or less as a closed system. Of course this structure is enriched by memorised experience. But in reality each sub-set has within itself the same finality as the set: the protection of its integrity within time. All humanity's unhappiness arises because we have not yet found a way of including this closed structure in the largest set, whose finality would be both its own and that of all other structures. Our unhappiness arises because we have not found a way of converting the individual regulation into a servomechanism which is included in the species. The organism is a system which is open within itself, in levels of organisation: it is a chain of servomechanisms. The entity it represents is open from the point of view of circulating information, since it is informed of what is happening in the environment through its sense organs. But the information collected is used by the organism only in order to act on the environment in the interest of preserving its information-structure: as the diagram shows (see figure 4), the loop originating in the environment closes upon the latter. The entity is also open in the thermodynamic sense: but its structure is closed.

The only way of opening the information-structure of an organism, of opening the regulated individual organic entity, is to convert it into a servomechanism, that is to say, to include it in a higher level of organisation, the social group, whose finality must however be the same as its own. Unfortunately the social group immediately becomes a closed system whose finality is to maintain its structure, and is obviously in opposition to the surrounding social groups unless these groups join together as sub-sets of a larger set. Once again, it is necessary to find a finality for this new set which is identical to that of the constituent sub-sets.

I am well aware of the unorthodoxy of this notion that the information-structure is closed; it seems difficult for many people to accept. There are several reasons for this. The first is that Shannon's theory does not deal with the semantics of the message. The second is that information cannot be separated from the mass and energy which convey it: the information-structure is supported

25

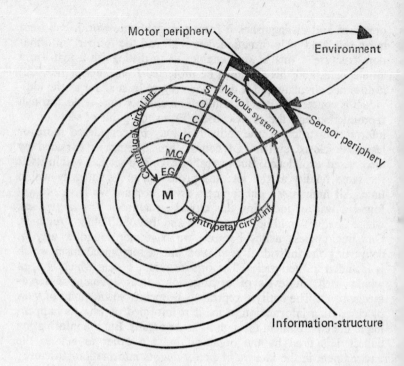

Centrifugal circul. inf. =
  circulating information, centrifugal
  from the nervous system

Centripetal circul. inf. =
  circulating information, centripetal
  towards the nervous system

M   = molecules
E.G   = enzymatic groups
M.C   = metabolic chains
I.C   = intracellular organelles
C   = cells
O   = organs
S   = systems
I   = individual

Figure 4

by elements and maintained by energy which circulate through the system, so it must be open to them. This is backed up by the fact that disorganisation of the information-structure is caused by the aleatory energy of the environment. It has been suggested to me that I should therefore describe the information-structure as "fixed", rather than "closed". But this would cut out the possibility of its achieving an opening by membership of an encompassing system whose finality it must then adopt. And if it does not achieve this, it remains a "closed set".

The notion of finality

This term does not imply finalism in the philosophical sense. Its meaning derives from the application of cybernetic laws. An effector, i.e. any mechanism enabling an action to be carried out (an effect), is directed towards a goal, for it has been programmed to achieve it. The eye is made in such a way that it takes part in the phenomenon of sight. Pittendrigh[4] replaces the term "finality" by "teleonomy", and this has been taken up by Monod to designate the action of systems working on the bases of a programme and of coded information.

An organism consists of structures which have a functional finality, and which through the levels of organisation contribute to the finality of the set — a finality which may be called "survival of the organism", and which is the result of preserving its complex structure in a less complex environment. This appears to be a constant evasion of the second law of thermodynamics, the law of entropy. This leads us to the idea that the finality of each element, of each sub-set or part of a living organism, co-operates in the finality of this organism; but also that retroactively the maintenance of the organism's whole structure, its finality, secures the finality of each of these elements and thus the preservation of their structure.

Experimental methodology

Once these basic principles have been set out, one notices immediately that the essential method of experimentation is to observe a level of organisation while suppressing its external control. Experimentation reduces the servomechanism to the rank of a regulator. It closes the system at a certain level of organisation. The enzymologist and the biochemist isolate the elements of an enzymatic reaction *in vitro;* the biologist isolates infracellular structures in order to study their activity in isolation from the cellular system to which they belong, or he studies the biochemical activity of an isolated tissue. The physiologist isolates part of an organ or a whole organ in order to study its behaviour, or focuses his attention on one system, for example the cardio-vascular or nervous system, in order to study one favoured criterion of activity.

It is regrettable that the clinician himself usually acts in just the same way, treating "a heart", "a stomach", "a liver", etc., which involves isolating it from the familial and socio-cultural context in which the organism lives. This attitude, which is viable in an experimental context, is obviously one of the causes of the frequent ineffectiveness of treatment which deals only with the organic lesion.

This experimental approach is necessary, for the information which reaches a level of organisation is multifactorial, and there are too many variables to grasp all at once. It is therefore essential to place the level of organisation under study in a stable environment whose principal characteristics can easily be controlled. A single factor can be varied at will, so as to observe the consequences of the variations at the level of organisation under consideration. It would obviously be dangerous to conclude, from facts observed in these conditions, that the same things happen when the level of organisation is in place in the organic situation. However, this is the only way to acquire a progressive knowledge of the complex mechanisms which animate living matter. It demands, as I understand it, a constant to-ing and fro-ing on the part of the experimenter from one level of organisation to another. In other words it demands an "openness" of mind that is capable of adapting itself to the "opening" of the complex systems formed of living matter.

It is important to understand this notion, for one often hears it said that the social, the economic or the political cannot be "reduced" to the biological: this is the fashionable battle against "reductionism", a battle in which I would be the first to take part. Indeed, for a long time I tried to display an interdisciplinary attitude in my work, at a time when this attitude was very unfavourably regarded and only the reductionist specialists were trusted. But while there is no question of reducing the functioning of the central nervous system to that of the isolated neuron, how can we understand the functioning of the nervous system without knowing how the neuron works? When doctors used to observe stiffness of the muscles in the nape of a patient's neck, headaches

28

and mental confusion, sometimes a coma, a high temperature, slow pulse-rate and vomiting, they would diagnose the meningeal syndrome (let us note that it had taken thousands of years to bring all these separate symptoms together and show that they expressed irritation of the meninges). But because until only a few years ago they limited themselves to treating the symptoms out of ignorance of the underlying processes, they prescribed ice-packs on the head and aspirins, and most patients died. Progress came from knowledge of what was happening in the levels of organisation situated beneath the symptoms. Since the technique of lumbar puncture enabled specimens of the cerebro-spinal fluid to be taken, and microscopic examination of the latter allowed the isolation of a germ or pathogen and the culture of certain of the germs responsible for the infection, the study of their sensitivity to certain antibiotics and the injection into the patient of the most effective of these for the germ concerned have enabled almost all patients to be cured. It is through the study and knowledge of phenomena situated at levels beneath the symptoms, through the understanding of the mechanisms by which these molecular, cellular and systemic phenomena cause the observed symptoms and no others, that effective treatment is arrived at. This does not mean that diagnosis does not always begin with the recognition and grouping of symptoms into a meaningful group, but it means as well that effective action begins only with a knowledge of the servomechanisms which link each level of organisation, from the molecule to the symptoms. This knowledge is almost always indispensable if the treatment is not to be merely symptomatic and thus ineffective, but etiological (causal) and effective. No level is an "encompassing" one if there is ignorance of the levels it encompasses.

Now sociology, economics and politics, and very often psychology, being enclosed in conscious speech (the "discourse on . . .") and cut off from the underlying levels of organisation, are behaving in a similar way to the doctor of a few decades ago who diagnosed a pathological disturbance. Progress in medicine during the last few years has come from progress in biochemistry. Without fear of being mistaken, we can prophesy that future progress in the human sciences will come from the recently achieved progress in biology, and particularly in behavioural biology. This is the opposite of the reductionist attitude, for the "reductionist" is someone who concerns himself with one single level of organisation.

I am not attempting, as I have often been accused of doing, to reduce the "psychological" or the "sociological" to the biological.

This implies a certain attitude on the part of the accusers: for them, the psychological is only the consciously psychological, and they themselves are already "reducing" psychology to the fringe of surface foam which covers the deep swell of their unconscious. If biology is defined as interest in the living world, then all the "human" sciences are biological. One cannot enclose oneself in a single level of organisation, whatever it is, unconsciously wishing to remain ignorant of the others in a highly primitive attitude of territorial defence and attempted repression of the unknown (i.e. of what is inevitably a source of anxiety). It is not the reduction of the psychological or the sociological to the biological that is to be feared, but the reduction of the sociological to the sociological, or of the psychological to language. What is to be feared, in other words, is the closing of a system of thought, an information-structure, upon itself.

Open systems and evolution

These two types of opening which are "actualised" at the thermodynamic level and the informational level are characterised by the fact that each regulated system, at whatever level of organisation it is situated, is at the same time part of an energy or information chain which opens out on to the levels of organisation immediately above and below it. It is interesting to consider the evolutionary process from these two different points of view.

In a regulated system the value of the factors is controlled by the value of the effect, by means of feedback. In this sense, an animal species in its ecological framework can be considered a regulated system. At the thermodynamic level, as long as it finds substrates in large enough quantities in the environment to maintain its "information", in other words its structure, and as long as its waste matter doesn't accumulate or can be reprocessed in the great cycles of the biosphere (oxygen, nitrogen, carbon, water, etc.), its specific equilibrium will be preserved. There is no obvious reason for this equilibrium to change. The ecological system seems to be closed at the level of the information-structure, and no evolution towards greater complexity seems possible. Now we know from observation, since Buffon, Lamarck and Darwin drew our attention to the subject, that the evolution of species does exist. Today many people find it necessary to interpret this evolution by attributing it to haphazard mutations occurring in the gene pool. I do not intend to undertake a criticism of this theory here, nor to expose the many objections it comes up against. Others have done this better than I could. I would simply like to draw attention to one aspect of the problem. We have just noted the existence of levels of organisation which are characteristic of living organisms, and the fact that these levels of organisation represent the elements of what I called their "complexity". We have also seen that, for a given species, the information of this complex structure is already complete in the genome. In terms of the theory of sets, it could be said that the genome represents a set of genes whose atomic and molecular structure is defined for a given species. A mutation would be a new organisation, establishing new relationships among the elements. Such a mutation might give birth to a new organic form. But it is difficult to imagine how it could cause evolution towards greater complexity without the addition of new elements to the genome. How, for example, could it account for the appearance of *additional* brain areas (the limbic areas) in the development from the simplified brain of the reptile or amphibian to the mammalian brain? How in turn could it account for the development and addition of the human orbito-frontal lobes to the corticalised mammalian brain? In short, even if some genes seem to have no purpose in the simplest species, one is inclined to think that evolution towards greater complexity must be the result of a union or intersection of genomic pools, rather than the mutation of one original pool. It is easier to envisage hybridisation and symbiosis than a chance favourable mutation which became stabilised because it was effective (that is, selected by the environment).

Certain recent ideas allow us to illustrate this opinion. The biochemical and structural similarities between bacteria and the intracellular micro-organisms known as mitochondria have led people to consider that mitochondria might once have been bacteria which colonised the cytoplasm of host cells.

Mitochondria do in fact possess a DNA which enables them to synthesise their own proteins without having to use the DNA of the cell nucleus. Yet it seems that the mitochondria have lost some genes because they are parasitic. There is no room here to sift through the huge amount of information about this problem which has accumulated during the last few years. However, it is important to know that when isolated mitochondria are placed in contact with a culture of fibroblasts, they are still capable of functioning. Now in my opinion this is where the problem becomes interesting. The function of mitochondria is essentially their ability to use oxygen to receive electrons at the end of biocatalytic chains. Once

31

the cascading electron has lost its excitation energy, it finds its base orbit on free radical oxygen molecules and atoms (i.e. those with unpaired electrons). In its course within the mitochondria, the energy which the electron releases can provide an amount of ATP which is relatively large compared to that provided by other, more primitive systems of storing intracellular chemical energy — these other systems are incapable of using oxygen which appears after them.

In fact the original atmosphere of the earth did not contain molecular oxygen, and life-forms appeared without it. The appearance of oxygen is a result of chlorophyll photosynthesis and is to some extent a by-product, as Boivin stated, of the photolysis of water. Before oxygen occurred, life-forms used organic molecules at the end of biocatalytic chains in order to receive electrons. At that time the process was fermentation, an anaerobic process which is capable of being carried out without molecular oxygen. These processes still form the essential biochemical mechanism of many life-forms. But the quantity of ATP synthesised by fermentation is much smaller than that which is synthesised by oxidising processes. The waste products are, for example, ethyl alcohol or lactic acid, which are very incompletely broken down. These compounds still contain much unused potential energy. Now these waste products can be used by mitochondria which serve as substrates to them, breaking them down into carbon dioxide and water. We can see the efficiency of the symbiosis between these primitive anaerobic forms (also known as glycolitic forms, because they break down carbohydrates) and the primitive aerobic bacteria which developed into mitochondria, using oxygen to pick up electrons: there is a considerable increase in the capacity for storing energy in the form of ATP.

Unfortunately oxygen is a biradical, and is toxic for the original life-forms. Indeed, while seeking to pair off its deficient outer orbit of electrons, it may capture electrons from the paired, stable molecules of biocatalytic chains and thus provoke forms with free radicals which will cause reactions removing electrons. In other words they will produce a chain reaction of oxidation, similar to ionising radiation. If this occurs, the disintegration of the fragile living structures will occur.

When living systems began, the anaerobic forms were protected from the sun's ultraviolet rays by the surface of the primitive oceans in which they were born. With the appearance of chlorophyll and photosynthesis, molecular oxygen appeared in the atmos-

32

phere and was condensed in the higher layers of the atmosphere into ozone, which served as a protective screen against ultraviolet radiation and enabled life-forms to emerge from the oceans. The birth of bacterial forms capable of using molecular oxygen and of stabilising it by pairing up its electron-deficient outer orbit is one example of the mechanisms by which the primitive evolution of the species was able to proceed. The symbiosis of bacterial forms with the more primitive anaerobic forms to produce most of the cells of contemporary organisms is another example.

I have dwelt at some length on these problems simply so that I can show how much importance the thermodynamic and informational opening probably has, both in the processes which exist in contemporary life-forms and in the historical processes through which the evolution of the species has passed. This evolution was possible only because regulated systems (i.e. those which were closed at the level of the information-structure) were supplied with their opening by an external control, the servomechanism.

2

The nervous system

We deal with the nervous system at this point because its functional activity in man leads to the phenomenon of reflective consciousness. For a long time this consciousness was not believed to be simply the expression of the dynamics of complex living structures; it was thought to belong to a sphere fundamentally different from that of mere matter. It is certainly our awareness of what is usually called the human "mind" which has upheld and still upholds the conviction of many people that the world of "mind" cannot be reduced to the world of "matter".

Our present knowledge about the nervous system,[5] or at least that which concerns the sub-anatomical levels of organisation, is very recent. The biochemistry of the central nervous system in particular is scarcely more than twenty years old. This means that only in the last few years has it been possible to construct models linking up the molecular, metabolic, cellular and functional levels of organisation of the large regions or nerve pathways: in other words, to locate the servomechanism controls and to understand the regulating systems. This dynamic ensemble leads on to the question of behaviour. But modes of behaviour occur in response to the surroundings. At the level of organisation of the individual, the stimulus comes from the physical and socio-cultural environment, and the response is an action on the environment.

The nervous system consists of cellular elements called neurons (see figure 5). They consist of a "body" or "soma", with extensions. The extensions usually carry the impulse, either from the periphery towards the soma, in which case they are "dendrites", or from the soma to the periphery, in which case they are "axons". In the first case, the impulse has a cellulipetal direction; in the second, a cellulifugal direction. To pass from one neuron to another the impulse must cross a "synapse". This is the point of contact between two neurons. It connects the ending of an axon swollen into a knob (called the presynaptic nerve-ending) to the contact surface of the following neuron, which may be a dendrite (axo-dendritic synapse) or the body of the neuron (axo-somatic

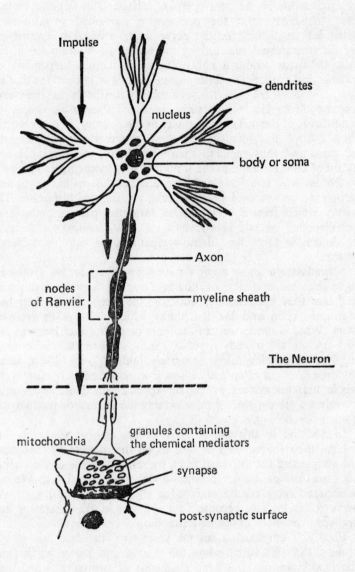

Impulse

dendrites

nucleus

body or soma

Axon

nodes
of Ranvier

myeline sheath

The Neuron

mitochondria

granules containing
the chemical mediators

synapse

post-synaptic surface

Figure 5

35

synapse): this is the post-synaptic surface. The impulse crosses the synapse because the presynaptic nerve-ending releases a "chemical mediator" for the nerve impulse which is synthesised by the neuron and released by the arrival of the impulse in the synaptic space, where it either excites or inhibits (depending on its chemical nature) the following neuron. It is not certain that all the substances believed to be chemical mediators of the nerve impulse do in fact act as such. Many of them may simply be modulators of the metabolic activity of the neurons; in other words, they control the intensity of these microscopic factories, the neurons, and thus regulate their excitability, the threshold and extent of their response to the stimuli of the environment. The main mediators are acetylcholine, adrenalin, noradrenalin, dopamine, serotonine, gamma-aminobutyric acid, glycine, and histamine. The nerves which release acetylcholine form the parasympathetic or "cholinergic" system. Those which release adrenalin, noradrenalin or dopamine form the adreno-sympathetic or catecholaminergic system.

Nowadays we know approximately how the neurons synthesise these chemical mediators of the nerve impulse, how they are stored and how they are released; we know the mechanics of their biochemical action and the machinery which destroys or recovers them. What is more, we can facilitate or inhibit each process. To sum up, we are already capable, through neuropsychopharmacology, of influencing fairly accurately (although the doctor using the drug does not often realise how it works) the functioning of the whole nervous system or certain specific functions, and consequently we are capable of transforming the normal or pathological human psychology.

In addition to this the nervous system acts on the organs to control their functioning, as well as on the muscles of the limbs, the vessels and the intestinal tract, by releasing some of the chemical mediators of the nerve impulse which we have listed. We can understand why recent knowledge of the biochemistry of the nervous system has become so important in the treatment and perhaps even more in the understanding of our behaviour.

Finally, I emphasised several years ago[6] the fact that in my opinion the cells surrounding the neuron and separating it from the blood vessels, known as glial cells or neuroglia, which had long been considered as mere supportive tissue, played a fundamental role in neurophysiology, and that the action of drugs affecting their metabolism could also affect the general working

36

of the nervous system. Neurophysiology in this case becomes the study not of the neuron but of the neurono-neuroglial couple.

The nervous system may be considered to have four basic functions (see figure 6): (1) The reception of energy variations occurring in the environment, via the sense organs. Its sensitivity will depend on their structure, and will vary from species to species.

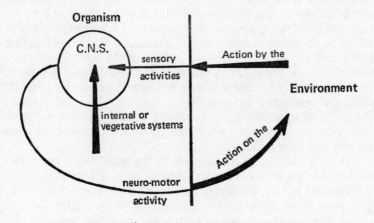

Functional outline of the nervous system

Figure 6

The dog and the dolphin can hear ultra-sound which is imperceptible to our ears. (2) Transmitting the information received to the higher centres, where there is also a flow of (3) internal signals, summarising the state of equilibrium or lack of equilibrium of the whole of the organism. For example, when the last meal was several hours ago, the resulting biological imbalance causes internal signals which, by stimulating certain lateral regions of the hypothalamus, will trigger off food-searching behaviour and, if the sense organs tell of the presence of prey in the environment, predatory behaviour. (4) This action on the environment, if successful, will permit a return to internal equilibrium and to the stimulation of other cellular groups in the same hypothalamic

37

region, causing the behaviour which is due to repletion. These types of behaviour, whose biochemical and neuro-physiological mechanisms are already extremely complicated, are among the simplest types and are essential for immediate survival, as are the mechanisms controlling the satisfaction of thirst and of reproduction, from courtship displays and copulation to the preparation of the nest, the early education of the offspring, etc. These forms of behaviour are the only ones which can be called "instinctive", for they carry out the programme which results from the very structure of the nervous system and are necessary for the survival of both the individual and the species. So they depend on a very primitive area of the brain which is common to all species endowed with higher nerve centres: the hypothalamus and the brain stem. When there is a stimulus in the environment, and the internal signal is present, these types of behaviour are stereotyped and cannot be adapted; they are unaffected by experience, for this simplified nervous system is capable of only a short-term memory, lasting no more than a few hours. These sorts of behaviour are a response to what we may call *basic needs*.

But each addition to the structure of the brain during the evolution of the species can be considered as derived from this primitive nervous system, modifying its functioning in increasingly complex ways as we climb the evolutionary ladder (see figure 7).

We must remember that originally the individual can find biological equilibrium, well-being and pleasure only by a motor action on the environment. This motor action in reality leads to the preservation of the complex structure of the organism in a less "organised" environment, by means of the energy exchanges maintained (within certain limits) between this environment and the organism. In contrast to this, plants' lack of a nervous system makes them totally dependent on their ecological niche.

The term equilibrium, which I am using to make the idea easier to understand, is a dangerous one, because it is difficult to use without implying a value judgement. It is customary nowadays to say that living systems are "open systems in a state of imbalance". I have already pointed out that although they are open thermodynamically, and open to what I have called circulating information, they are not open as far as their information-structure is concerned (see figure 4). As far as the notion of equilibrium in biology is concerned, I have said elsewhere[7] that in my opinion equilibrium is to be found in homogeneity and in death. Homeostasis, a term which is supposed to express equilibrium, does not

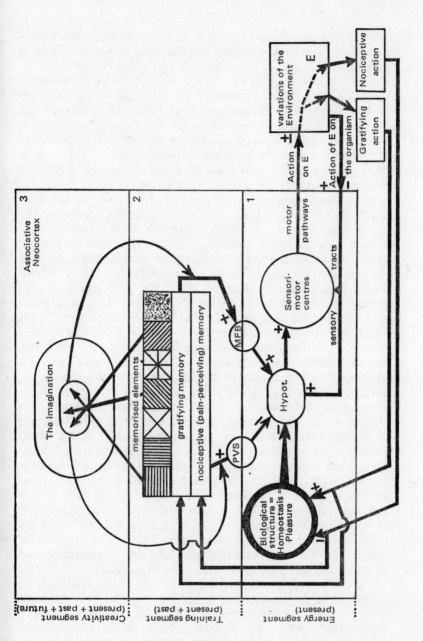

Figure 7

39

exist, except as an ever-unsatisfied quest. On the other hand, what does exist is a complex system of reactions, mutual adjustments programmed in the structure in such a way that this information-structure is preserved. By equilibrium we really mean the preservation of the information-structure, whatever means — whether more or less direct, more or less apparent — are used to achieve this finality. In this case every tendency of the living system towards entropy and the collapse of its structure without any compensatory metabolic activity can be called imbalance; but in reality it is merely a step on the path towards maximum entropy, death, and thus towards equilibrium and homogenisation.

In the first mammals, new formations appear as "derivations" from the preceding system — a feature generally known as the limbic system (McLean, 1949).[8] Although it is classically considered to be the system controlling affectivity, it seems to me more exact to say that it plays an essential part in the establishment of the long-term memory (Milner, Corkin and Teuber, 1968),[9] without which affectivity hardly seems possible.[10] The long-term memory is increasingly considered to be linked to the synthesis of proteins at the level of the synapses that are activated by experience (Hyden and Lange, 1968);[11] it is necessary in order to know that a situation has previously been experienced as pleasant or unpleasant, and in order that what is known as an "affect" may be set in motion by its appearance, or by the appearance of any situation which cannot be classified *a priori* as pleasant or unpleasant because of an "informational deficit" in this respect. Originally, the pleasant experience is one which entails a return to or maintenance of biological equilibrium; the unpleasant experience is one which endangers this equilibrium and therefore its survival, the preservation of the organic structure in a given environment. Long-term memory therefore allows the pleasant experience to be repeated and the unpleasant experience to be avoided. Above all it permits the temporal and spatial association, within the synaptic pathways, of memorised traces linked to a meaningful signal concerning the experience; and thus it causes the appearance of *conditioned reflexes,* both of the Pavlovian type (affective or vegetative) and of the Skinnerian[12] type (with neuromotor expression).

The synthesis of protein molecules following the stimulation from variations in the environment appears to "encode" the synapses through which the nerve impulse has passed. The nerve pathway used by the impulse is thus transformed more or less definitively, so that an analogous stimulation would tend to activate only the same nerve pathways and the same synapses which were activated by the first stimulation. If protein synthesis is pharmacologically prevented, the long-term memory is inhibited. We have recently shown that a medium-term memory also exists, whose protein synthesis is mitochondrial in origin (H. Laborit, B. Calvino and N. Vallette, 1973).[13] This synthesis seems to occur in a mitochondria-rich region immediately next to the synapse. We have seen that these mitochondria possess their own DNA and carry out part of their own protein synthesis without recourse to the nuclear genome (see p. 31). The short-term memory is no more than the consequence of a momentary persistence of the nerve impulse, started by an event in the environment as a result of reverberating circuits (especially reticular ones) which delay and prolong the effect. "Re-memorisation" or remembering would require the functional reactivation of the short-term memory, or more precisely of its anatomo-functional substratum, in order to reactivate the long-term memory.

Two bundles were discovered by Olds and Milner (1954):[14] the medial forebrain bundle (MFB), which we may call the reward and reinforcement bundle, and the periventricular system (PVS) or punishment bundle. By uniting various regions — the hypothalamic, the limbic and, in higher animals, the cortical — these bundles enable the cerebral unit to function effectively to satisfy instinctive drives, or to avoid unpleasant or dangerous experiences. The MFB is catecholaminergic and seems to activate the hippocampus, whereas the PVS is cholinargic and activates the amygdal (Margules and Stein, 1967).[15] The former appears to exert an inhibiting effect on the latter (see figure 7).

But on the other hand, by allowing automatisms to be created, the memory is at the origin of new needs which can no longer be called instinctive but are more often than not socio-cultural. These *acquired needs* become necessary for well-being and biological equilibrium, since they transform the environment or human action upon it in such a way that less energy then becomes necessary to maintain homeostasis. The result is a smaller range of reactions and a progressive loss of *drive*, i.e. a reduction of the margins of physico-chemical variations of the environment in which an organism can maintain its biological constants. These acquired needs may be the cause of hypothalamic drives which attempt to satisfy them by means of an act of gratification on the environment, but they may also come into conflict with other automatisms of a socio-cultural origin, which will prevent their expression. We can thus define need as the *quantity of energy or information necessary to maintain a nervous structure, whether innate or acquired.* The structure in the latter case is a result of interneuronal relationships established by training. It is certainly a "material" structure, since the neuronal relationships are established by means of the stable transformation of the synapses through protein synthesis, which is triggered off at their level by impulses caused by the external event. The *need* then becomes the source of the *motivation.* But because, as we shall see, in the social situation these needs can usually be fulfilled only by domination, the basic motivation in all species is expressed by the pursuit of domination. Hence the appearance of hierarchies, and of most of the subconscious conflicts which form the basis of what is sometimes called "cortico-visceral pathology", and which should more accurately be called "sub-cortico-visceral". But in the case of human beings, socio-cultural inhibitions and needs are expressed, institutionalised and transmitted through speech; the cortex is therefore also involved in the generation of a logical discourse expressing the mechanisms of conflict in the regions underlying it.

In the creatures which have evolved furthest, the existence of a *cerebral cortex,* which in man has well-developed orbito-frontal regions, provides a means of associating the memorised elements. Indeed, since these elements are incorporated into our nervous system through different sensory canals, they are linked in our long-term memory only because our action on the environment shows through experience that they are linked in a certain order, i.e. that of the perceptible structure of an object. But if sufficiently developed associative systems such as those characteristic of the human orbito-frontal lobes are capable of recombining these memorised elements in a different way from that in which they are received from the surroundings, then the brain can create new structures, imaginary structures. The new-born child can imagine nothing, for it has memorised nothing; the imagination tends to be richer as memorised material is more abundant, provided that this material is not imprisoned in acquired automatisms. Indeed, along with speech, which gives us access to concepts and enables us to stand aloof from the object, humanity's ability to manipulate abstractions by means of the associative systems gives it almost infinite possibilities for creation (see figure 7).

The nervous system as I have just outlined it is termed the "life" of relationships, because it brings the organism into a sensorimotor "relationship" with its environment. This sensorimotor arc permits the adjustment of the environment to the homeostatic internal equilibrium, by means of *action* on this environment. But in order to carry out this motor activity a large amount of energy must sometimes be expended by the locomotor organs, the muscles, and by the organs which ensure their blood supply and which must also have preferential vascularisation, as well as the cerebral and spinal centres which control the muscles. This is the case when struggle or flight are the only types of behaviour which can ensure survival. The "vegetative" nervous system, one of whose principal roles is to carry out the vasomotor adjustments necessary for the preferential supply of the preceding organs to permit flight and struggle, also permits — with the help of its neuromodulators and of medullo-suprarenal secretion — the metabolic adjustments on which the release of energy and the functional activities are based. The same applies to the endocrine system, whose hypophysary control in turn depends on the hypothalamus.

Here again we encounter the notion of homeostasis. For a long time we have had to distinguish between a homeostasis restricted to the internal environment in which the information-structure of the organism's cellular set lies, and a generalised homeostasis of this organic set which may sometimes require the loss of the restricted homeostasis. The latter occurs when the flight or struggle reflex is set in motion to safeguard the general information-structure; this necessitates, as we saw earlier, a temporary sacrifice of the energy supply to certain organs which are not directly involved in this behaviour. If this reaction lasts, because it has failed to remove the danger which threatens, the information-structure itself may suffer, and a state of shock and then death occur. So the internal medium, a liquid cushion mediating between the variations in the environment and the cellular information-structure, acts as a passage for the matter and energy which the information-structure needs in order to survive. It also acts as a passage for the waste products before they are excreted into the environment. Its composition is thus constantly disturbed by the variations which spring up in the environment. In the *psychological state*, the cellular information-structure keeps its composition constant by countless feedbacks and servomechanism loops. In a *state of emergency* this constant state is temporarily sacrificed for flight or struggle. The constant composition may be re-established if danger is averted by a victorious struggle or flight. If the state of emergency persists, a pathological state may be reached in which the information-structure itself is damaged, either acutely or more gradually, with

43

the eventual appearance of chronic lesions which are most likely to be found in the organs sacrificed by the organic reaction to aggression. Homeostasis can therefore no longer be considered as the tendency to maintain "the conditions of life in the internal medium". It is the tendency to preserve the integrity of the organism's information-structure — sometimes by keeping the conditions of life in the internal medium constant, and sometimes by means of the organic unit's motor autonomy within the environment but at the expense of the constant conditions of life in the internal medium. So the finality remains the same: the preservation of the structure. But the programme used to achieve this may change, as may the methods used. The notion of homeostasis, a word suggested by Cannon[16] on the basis of a concept of Claude Bernard's,[17] is of considerable importance in contemporary physiological thinking. But it has unfortunately rigidified this thinking by very often preventing it from looking any further for the key to effective treatments. For many years we have limited treatment to the revival of the internal medium, and only very belatedly have we become interested in reviving the information-structure, which is dependent on the correct functioning of the cellular chemical factories, which in turn supports the correct functioning of the organs and systems. This correct functioning is responsible for keeping constant the conditions of life in the internal medium.

Conscious states

A state of consciousness obviously requires some notion of the bodily outline. The new-born baby enclosed in its "ego-whole" can probably not be described as conscious even when awake, for it has not yet become aware of an image of itself separate from the outside world. To do this, it must experience a motor action on the environment combined with interoception. We know how important the hypothalamic nucleus, which integrates the pulvinar, is in establishing this outline. It is now accepted that in the child the functional activity of the sensory pathways from different sensory canals links up in successive stages with sensory discrimination in time and space. A state of consciousness must also be related to an experience of oneself in time and thus requires memory processes, for consciousness is primarily a consciousness of the bodily outline lasting in time. To be conscious one must be awake, in order to compare constantly present stimuli with the experience of past stimuli. An experimental state of sensory deprivation rapidly leads to sleep and loss of consciousness. Memory processes and motivation depend on the limbic system and hypothalamus; wakefulness depends on the activation of the reticular formation of the brain stem. But we know that any reflex or automatic action is usually unconscious. This even seems to be its main usefulness, for it releases the system which concentrates the attention (the diffuse thalamic system) while enabling the action to be carried out. This is the advantage of all sorts of habits which are in the nature of a "craft" and depend on "training" (in society too, it is necessary to have training, i.e. an apprenticeship, in order to obtain a craft); and we know the role of the limbic system in the training and memory processes. By themselves, therefore, these processes are not enough to produce the phenomenon of consciousness.

Conversely, a strictly aleatory mode of behaviour which is unpredictable (except in terms of statistics) implies the absence of memory: the response of the nervous system to variations in the environment would be different each time, since a situation never reproduces itself. This behaviour would also be unconscious: consciousness is primarily the consciousness of the duration of the individual in time and is thus a function of memory, especially memory of the unity and duration in time of the individual subject which remembers the variations that have occurred in the environment and its own dynamic relationships with them.

So consciousness seems ultimately to result from the fact that it is impossible for a neurophysiologically and ideally normal individual to be unconscious: that is, to respond by either completely automatic or completely aleatory behaviour. The fact that the memory and its experience, whether innate or acquired, causes the individual to respond automatically — that is, unconsciously — is borne out by his whole sub-cortical and non-associative cortical system and essentially by his paleocephalus, which he shares with the other animal species. Speech alters nothing; it is simply an extra and more complex means of stimulation, a second system of signalling (Pavlov) which is capable of enriching his behaviour without necessarily making it more conscious. The associative systems, however, cannot be envisaged in isolation from the preceding systems, since if they remembered nothing they would have nothing to associate. But if they could associate the memorised elements at random, with no relationships between them, and particularly with none of the temporal relationships which are necessary for the subjects' awareness, awareness that he will last

45

in time, then this function will of course be unconscious. In the final analysis, man is conscious because he responds in an original way to the problems posed by the environment, whereas he could respond in a reflex or automatic way. Therefore, he should become more conscious to the extent that he is conscious of his automatisms and drives, and to the extent that he is capable of freeing himself from them by his imagination. It might also be thought that he would tend to become more conscious as his fundamental drives are more powerfully contradicted by the socio-cultural taboos which the automatisms create. But most often this insoluble conflict is so painful (the PVS system) that the individual prefers to bury it in his unconscious, to repress it. (Besides its associative role, the cortex also possesses an inhibitory effect on the underlying areas.) No doubt this is one of the essential sources of neurosis and psychosis.

Consciousness is thus shown to ensue from the most complete and most integrated functioning of all the cerebral regions and functions.[18] The animal thus becomes more conscious the less it is subject to unconscious automatisms. This should be the characteristic of the human species.

Psychosocial aggression

Many organic reactions can be shown in the simple act of a cat faced with a dog: for example, the blood circulation in the surface part of the kidney stops (Trueta, 1948).[19] The notions of "stress"[20] and "alarm reaction"[21] put forward by Selye have drawn attention to the a-specific nature of this reaction: "the organism is not a polyglot — it always replies to the questions posed by the environment in the same language". What distinguishes physical aggression from psychosocial aggression is primarily the absence in the latter case of any lesion directly traceable to the aggressor. The lesion resulting from psychosocial aggression is secondary to the a-specific reaction.[22] This is characteristic of psychosomatic pathology. What, then, is the link between stimulus and response?

If we accept that the peripheral adreno-sympathetic reaction, which is an essential element of the response, is programmed to maintain our motor autonomy from the environment by authorising flight or struggle, as we have just seen, then the stimulus must rely on memorised experience to stimulate this reaction. Psychosocial aggression, therefore, seems to activate the limbic system and the memory. In this way the limbic system may be considered

46

to control affectivity. The consciousness we have of affects is, moreover, essentially an awareness of the acute cutaneous or visceral vasomotor responses which accompany them. I have already explained why. It seems difficult to talk of psychological aggression in the new-born human being, which is unable to move around independently in the environment and does not yet have any memorised experience of the environment concerning what may threaten or encourage its survival (or pleasure). In this case the aggression will really be "physical", either because of direct action by the aggressor or because of the "lack" of an element which is essential for biological equilibrium (an alimentary factor, for example), and it will be fixed in what we may call an "ego-whole" since the bodily outline has not yet formed. Until it is formed, at the age of about eight to ten months, the baby will gradually build itself a memory of what is pleasant and unpleasant. This is linked with its environmental niche (in which the mother obviously has pride of place), but the baby cannot relate the experience later to an image of itself which is distinct from the outside world. Indeed, a state of consciousness must be related to an experience of oneself in time, and thus to a memory process which is extero- and intero-ceptive, for consciousness is primarily awareness of the bodily outline lasting in time. This memory engrammed in the "ego-whole" may later be the cause of phantasms, which are all the more distressing since it is then impossible to relate them to a past experience. It seems to me that this is the basis in animals of what Konrad Lorenz has described as the "imprinting" phenomenon.

Anxiety and distress

The two terms anxiety and distress express very similar sensations, differing only in their degree of intensity. Originally we can satisfy our search for biological equilibrium, well-being, pleasure and survival in our environment only by a motor action on the environment. It follows that any obstacle to this active gratifying behaviour, or especially any taboo against it, may on the one hand trigger the a-specific (mainly vasomotor) reaction whose finality is to permit flight or struggle, and on the other hand may cause a feeling of motor impotence. The combination of these two elements seems to lead to the sensation of anxiety or distress, for since its mechanisms are unconscious, this fear apparently has no legitimate object and causes a distressing feeling of expectancy.

Anxiety and distress may be reabsorbed in several ways. Let us consider the causative mechanisms first, before considering the escape mechanisms.

It appears obvious that in the social situation the hypothalamic drive (the Freudian Id), the individual's pursuit of pleasure, will collide with that of others. In all animal species this factor is at the origin of hierarchies and of the establishment of domination. In primates as in man, observation shows that the offspring of dominant subjects more often than not become dominant themselves, as a result of the education they receive. But in man, the rules to follow for establishing domination are institutionalised and transmitted through several generations by means of languages, and they form the essentials of a culture. If laws represent the socio-cultural taboos which are valid for all citizens in theory, in practice the taboos seem more numerous and oppressive the lower one goes down the hierarchical and economic scale.

Conflicts between instinctive (particularly sexual) drives and socio-cultural taboos (the Freudian Superego) are one of the primary sources of distress. It is important to emphasise that from the moment of birth, the individual is caught in a socio-cultural framework whose essential aim is to create automatisms of action and thought in him which are indispensable for preserving the hierarchical structure of the society he belongs to. Automatisms of thought form the value judgements and prejudices in general of a society and an era. Automatisms imply unconsciousness, and indeed we are unconscious of the socio-culturally determined nature of almost all our judgements. Since we are also unconscious of the biological significance of our drives, the conflict between the two usually remains in the unconscious.

If distress can be reabsorbed in action, conscious discourse will always provide an alibi, a logical analysis justifying the ensuing behaviour. But it should be noted that while hierarchies cause situations of conflict and distress, they are also a source of security. The creation of conceptual and behavioural automatisms of socio-cultural origin enables existential anxiety to be concealed behind simple explanatory grids; "chiefs" take over responsibility and give one a feeling of security, while people less fortunate than oneself can be patronised and thus satisfy one's congenital narcissism. Unfortunately this castrates all creativity, by punishing any project which does not conform to the system of values imposed by the dominant. Unconditional surrender to this system considerably limits gratifying action, yet it unconsciously mobilises an

48

organic reaction because of the partial dissatisfaction which ensues. This is probably a major cause of the so-called "psychosomatic" ailments. Indeed, we have already noted that the reaction permitting flight or struggle creates circulatory and nutritional disturbances in the organs which are not immediately essential to this behaviour; in the long term, lesions will result in these organs, with a gradual loss of their normal information-structure.

Another causal mechanism of distress arises from what may be called an "informational deficit" concerning an event occurring in the environment. Training [*apprentissage*] in the pleasant and the unpleasant allows us to decide whether to classify a new event or not. If it can be classified, active reinforcement or evasion behaviour will avoid distress. If it cannot, the impossibility of acting effectively again causes distress, just as the impossibility of acting faced with a dangerous but inevitable event. Let us also note that the anxiety-causing unknown is not always an event, but is very frequently constituted by the other, and by uncertainty about the other's behaviour.

In man the existence of the imagination, whose raw material is everything (conscious and unconscious) that has been memorised, is perhaps the most frequent cause of distress. Indeed, training in the existence of different forms of displeasure and pain provides a raw material whose plasticity is easily manipulated by association; new imaginary structures are created which may never come into being, but the fear that they may happen can inhibit action. They may come into synaptic conflict with the previous factors, the hypothalamic drives and socio-cultural automatisms, the pursuit of pleasure or the avoidance of punishment (MFB). Punishment, if not avoided, may bring the PVS into action. We shall see later how these different mechanisms of distress in the individual seem to be the cause of what is generally called "social unrest".

On this outline basis of the main mechanisms of distress, we can now deal with the ways of avoiding psychosocial aggression as I previously defined it.

Aggression

If we agree that distress usually occurs when it is impossible to achieve gratification, then the simplest, most basic reaction to escape from distress seems to be aggression. Delgado (1967),[23] and Plotnik, Mir and Delgado (1971),[24] have shown that aggression in the chimpanzee is closely connected with the existence of a painful stimulus and that it is directed at the dominated animal, not the dominant one. In other words, the uncontrolled explosion of aggression is a way of escaping the distress which is inevitable when gratification cannot be achieved. It may be produced experimentally by stimulation of the PVS, or inhibited by stimulation of the MFB. Aggression is a simplistic way of resolving the conflict between the hypothalamic drives and the socio-cultural taboos resulting from training. The dominant animals are not aggressive once they have established their domination, since this allows them to satisfy their pursuit of pleasure. The quantity of urinary catecholamines and their metabolites, along with the quantity of suprarenal corticoids, shows how weakly their neuro-endocrine system is stimulated in comparison with that of the dominated animals (Welch and Welch, 1971).[25] Conversely, the characteristic cerebral biochemistry of the dominant animal which has been victorious in the struggle is an extra amount of catecholamines, whose important role in the functioning of the MFB has already been indicated.

I am therefore obliged, from my experimental study of aggression, to oppose the widespread view of the last few years that man is inherently and congenitally aggressive.

The term "aggression" is dangerous, for its conscious expression in the human race gives the word a semantic content which behavioural biochemistry and neurophysiology cannot accept as an all-embracing label for a way of behaviour. Those who talk most often about "aggression" seem never to have taken the trouble to define it precisely. I have defined it elsewhere[26] as the quantity of energy capable of increasing the entropy of an organised system, its tendency to level out thermodynamically, in other words to make its structure disappear more or less completely. In this sense the predator is aggressive towards its prey without feeling either hatred or malevolence towards it. The predator simply supplies the cells of its organism, its chemical factories, with substrates so that they can preserve their structures. In this sense we might also say that all living structures — even the simplest — which preserve their structures in a less organised environment suffer constant aggression from it, and if they do not surrender to it, then they behave aggressively towards the environment in return.[27] In this case life and aggression are synonymous.

What is the connection between this behaviour for immediate survival and human aggression? There is a relation between, on the one hand, the instinctive behaviour that secures the satisfaction of the basic needs without which life is impossible, and, on the other, a training in environments which endanger survival and generate suffering that cannot be avoided by flight or struggle. But there is also a training in social "values", varying with the time and place; these are automatised from childhood within the human nervous systems, and their finality is the protection neither of the individual nor of the species, but of the social organisation, of a type of hierarchy in which dominant and dominated will always exist. Because the dominated are unlikely to try to keep a hierarchical system stable, then very early on the dominant must instal in the nervous system of all the individuals in the group a type of socio-cultural automatisms and value judgements which will help to maintain their domination and the hierarchical organisation of the group. The propagation of the idea of an an innate aggressiveness in humans, which we owe to the animal species which have preceded us in the phylum, is surely itself a part of this training process. Indeed, by creating a conceptual automatism concerning the animal origin of our aggressiveness man, who thinks he is so distinct from the animals, is obliged merely to control his aggression and so to respect the value hierarchies. It may be said that the aim of all socio-cultural training is the creation of submissive automatisms towards hierarchies, and the actual elimination of a type of aggression which is not in the least "animal" but arises when the individual cannot achieve gratification. In fact, apart from a type of aggression we may call "instinctive" (that is, a type of behaviour which fulfils the basic needs without hatred, and which in animals never leads to murder within the species), human aggression is never anything but a way of resolving distress. Distress, as we have seen, usually requires a training in pain; it is the result of a conflict between neuronal activities based on different levels of organisation in the human nervous system, a conflict which cannot be resolved by means of an act of gratification on the environment. The iron yoke of taboos, hierarchies and social structures which inhibits all gratifying activity is nowhere heavier than in contemporary urban and industrialised societies, since flight here is impossible. No doubt this is why apparently wanton aggression and violence are most frequently found within these societies, which are governed by production alone, and where the human being is nothing but a commodity-producing machine. This aggression and violence is accompanied by what is called the "malaise" of living. The lack of satisfaction from such a life-style even removes all fear of punishment. It seems to be in this same context that

51

alcoholism, suicide and drugs attract large numbers of individuals seeking relief from their frustrated pursuit of well-being, pleasure and joy.

It is worth noting here the different emotional charge in each of these words: well-being, joy, pleasure. Well-being is acceptable, joy is noble, pleasure is suspicious. The last word smacks of fire and brimstone. Whereas well-being appears when the drive or acquired needs are satisfied and is accompanied by satiety, and whereas joy seems to involve the imagination in this satisfaction, pleasure is bound to the present time, to the accomplishment of the action. It is no dirtier, no uglier, no less moral than the other two. It is clear that the different meanings given to these words are due to social and cultural automatisms, to value judgements which arise mainly from the sexual repression which has attacked Western societies for thousands of years, and whose main cause was probably fear of the unknown bastard who might benefit from the inheritance of private property.

Let us also note how often the reference to "nature" or the "natural" is used as an excuse for the value judgements of the time. One resorts to nature to prove that aggression is inevitable in man since it exists in animals. This justifies hierarchies and domination, the aggression of the dominant individuals in response to the aggression of those dominated (though of course it does not justify the aggression of those among the latter who "act like wild animals"), and wars. But incest, free love, and generally speaking everything that concerns sex (and consequently private property) are habitually practised without complexes by animals, and therefore they "lower" man to the level of animals. Reference to what is "natural" is simply an excuse for defending the dominant ideology. In the human species, language has allowed the rules of domination to be institutionalised, transmitted through the generations and linked not to the individual but to the social group. Crude, explosive violence (the only sort which is ever talked about) is simply a response to a stimulus; and this stimulus is nothing but institutionalised violence which, because it is institutionalised, no longer recognises that it is violence. At an earlier stage in history, this violence has enabled a socio-cultural group to become dominant. Certainly the violence of the Revolution and the Terror, institutionalised by the immortal principles of 1789 for the benefit of the bourgeoisie, is what enables the bourgeoisie to stigmatise the violence of the dominated today, with right apparently on its side.

I should also like to take advantage of the fact that we are talking of aggression to discuss briefly the notion of territory. The ethologists have taught us that this notion exists in animals, who "defend their territory". In this case it is not surprising that man should do the same. It must be a normal feeling, since it exists in animals. It provides a legitimate excuse for private property and patriotism. Man is an animal, all animals defend their territory, therefore it is just and fair that man should defend what is his, without seeking to know where the notion of property comes from. All the same, it must be noted that as yet no one has shown any cell group or differentiated nerve pathways in the hypothalamus or elsewhere which are related to the notion of territory and the behaviour connected with its defence. No territorial centre seems to exist. However, any gratifying behaviour must certainly occur within a certain "space" (we ought even to say "certain spaces", in the plural). Indeed, around the bodily outline a visual space, a sound space, and an osmic space are established; their boundaries are not the same, since they vary according to the acuteness of the senses of the species, and according to its possibilities of movement. Acts of gratification occur within these species: first, the search for food (predatory behaviour), and for reproduction (a mate). It is as if the individual were surrounded by a "bubble" whose boundaries are those of the acuteness of its senses, a bubble in which it moves and acts to preserve its structure, what we have called its biological equilibrium. If another individual opposes these acts of gratification, the first individual will become aggressive towards it. Thus territory becomes the "living space" necessary for the act of gratification to be achieved. It does not seem useful to talk of a particular "territorial" instinct. It is worth noting too that the extent of this space may vary according to the training in and satisfaction of basic needs. It follows that in the animal, territory is not "property" but a response to a need, which is more than just a never-satisfied need for conquest. Territory is rarely permanent throughout the life of the animal, and usually lasts only for the mating season. The "bubble", the territory, thus represents the bit of space which is in immediate contact with the organism, the space in which it "opens up" its thermodynamic exchanges in order to preserve its own structure. But since neolithic times, the growing interdependence of human beings as a result of the specialisation of labour has caused each individual's bubble to shrink, or at least has caused it to mingle with others to such an extent that we have become more and more concerned with the communal

bubbles of human groups. After the family, corporatism, region-
alism and patriotism should come humanism — not the humanism
listed in the encyclopedia, but that of humanity in its planetary
setting. But in contrast to this, with the growing promiscuity characteristic of modern cities, the individual's bubble has shrunk considerably. In residential areas, overpopulation and the invasion of
the acoustic bubble impose considerable limits on the space within
which acts of gratification can be carried out. If in other respects
radio, television and the press widen the visual, auditive and cogni-
tive bubbles, they do not at the same time facilitate action. On the
contrary, action becomes more and more stereotyped, less and less
effective in its basic aim of transforming the environment in the
best interests of survival. For thousands of years the individual saw
his place in the environment as a very limited space, but a space
in which he could nevertheless act effectively. Those events over
which he had no "hold" were the responsibility of the gods. Today
he lives in a planetary niche, even a cosmic one, and the gods are
dead. But he feels stifled by automatisms, by the intricacy of the
bubbles, by interdependence and the uncompromising surrender
to a social determinism which does not allow him to gratify himself
as he would wish. The economic machine grinds him down, and
he cannot resist or defend himself because it is so impersonal, so
infinitely variable and abstract.

Thus what is called territory is indeed the space in which an
individual may act to gratify himself. But in the same space there
are others who will limit the variety of these acts of gratification.
One of the problems faced by modern man lies in the fact that this
space is no longer a real space but a symbolic one, a considerably
magnified image: the others are very real, and are always there, and
encroaching further into the bubble in which he can act. Without
anyone asking his opinion he has got a pay slip and a national
insurance card — but he has lost the birdsong. The extent of his
territory depends on his rank in the hierarchy. The leader has a
much vaster territory than the skilled worker. The space in which
the worker can gratify himself is highly restricted. So restricted
is it that it is often simply the space needed for two bodies making
love, the only space which is still free — in certain cases — from
economic determinism and social prejudice. So we can see how
devious is the route which Mars and Eros, hand in hand, are forced
to take in order to reach the "territory" spoken of by the ethologists.

Linked to the notion of territory, understood in this way, is the
notion of property. Within an individual's territory, the space where

he acts to gratify himself, there are beings and objects. We know that gratification leads to a repetition of the act of gratification. Hence the appearance from childhood onwards of a tight bond between the object and the nervous system, the appearance of what is known as the "property instinct". Obviously this is not an instinct according to our definition of the word, but a type of behaviour which results from training, a training in gratification. It seems to me important to make this clear, because once it has been understood, the relationships between the notion of the ownership of people and things and hierarchical systems of domination can be explained simply, with no need to invoke the "essentially innate nature" of the ensuing behaviour. One wishes to own only those objects and people which enable one to achieve acts of gratification, and in particular their "reinforcement" or repetition. Property is like a drug: it is a habit-forming addictive poison which works by a cerebral process akin to compulsive addiction. In both cases the process is accompanied by the same synthesis of cerebral proteins that controls the stabilisation of every training process.

The notion of property is certainly the result of a socio-cultural training, since one may gratify oneself with collective goods (nature, the Parthenon, Beethoven's Fifth etc.) without trying to "appropriate" them. However, property is certainly linked to gratification, for even in these cases people will try to acquire a second house in the country and fill it with colour reproductions of their favourite buildings or pictures, or recordings of the music they love, so that the gratification is "reinforced".

A young child will easily let others have the toy he is not interested in, but he will scream if you try to make him give away the game he is playing with. The object of gratification which is in the immediate bubble of the child, even before the bodily outline has been established, is the mother. When he realises that she is also the source of gratification for his father, brother or sister, the child fears the loss of gratification and this may be when he unconsciously discovers the Oedipus complex, jealousy and unhappy love.

To sum up, these fundamental physiological bases are adopted by the adult who, in total ignorance of how trivial their mechanism is, elevates them at the social level into "ethics", into inalienable rights of the human being. Very early in life this behaviour is "reinforced" in the child by the adult, either by direct and indirect reward or by punishment, depending on the moral, ethical or legal rules of the society concerned. This is the origin of all value judgements and frustrations, which are all the more alienating because so far the only explanations supplied for them have come from a logical discourse based on false premises, and because they have had to be accepted as postulates relating to "human nature".

Flight

A comparison of the social life of modern man with that of his neolithic ancestors shows that certain ways of fighting or fleeing are impossible for him. In a natural environment, when two animals of the same or different species compete with one another either over territory or over a female, one of them finally gives up and goes away: this is known as "mutual agreement on an avoidance reaction", or mutual avoidance (Stephen Boyden, 1969).[28] The phenomenon is widespread among gorillas (G. Schaller, 1963).[29] When animals cannot avoid one another, for example when they are in a cage, competition often leads to the death or submission of the defeated animal. A hierarchy is established. In humans the same phenomenon appears. In primitive tribes "mutual avoidance" was still possible, and the comings and goings of individuals or groups can still be seen among the Bushmen (E. M. Thomas, 1959).[30] This has become impossible in our modern societies. Workplaces, the city, and the family home are meeting places between individuals where promiscuity is inevitable and where economic dependence creates bonds of submission which make the "mutual avoidance reaction" impracticable. We are in a cage similar to that of the two gorillas. The relations of production are not the only grounds for antagonism which can occur in this case, and relations of domination can themselves be a reason for "mutual avoidance".

Depression

When no possible escape from psychosocial aggression exists, a depressive state is often encountered. Many etiological forms of depression have been described. Depression is chiefly the result of being unable to satisfy the pursuit of pleasure through some act of gratification. But this pleasure may be instinctive, it may be hypothalamic, or it may be the result of a training process, of acquired socio-cultural automatisms. The loss of someone close (a cause of bereavement depression) is of the second type, because there is a brutal interruption of the reinforcement of ritualised interindividual relationships and a breaking of interpersonal bonds.

The central depletion of catecholamines, which is generally noted in a depressive state, can be interpreted by the inability to perform an act of gratification on the environment because its object is no longer there, and thus an inability to set the catecholaminergic MFB in action. Similarly, depressive states can be caused by the use of drugs inhibiting the activity of the catecholaminergic neurons, such as reserpine, 6-hydroxydopamine, or α-methyltyrosine. We can also explain the anti-depressant activity of amphetamines, IMAO or tricyclic anti-depressant compounds in this way. But the various organic lesions, which are secondary to the vasomotor and endocrine reactions constituting the field of psychosomatic pathology, are undoubtedly a way of expressing so-called "moral" pain, in other words the subject's inability to use his MFB activity to trigger the control of his PVS activity by means of a gratifying form of behaviour. We should also point in this context to compensatory behaviour: for example overeating, in which there is a transference to food of the gratifying behaviour which has been forbidden elsewhere, in some unconscious drive or acquired automatism.

Addiction

In this general model, another means of escaping distress appears to be addiction. The addict (unless he is an alcoholic) is a non-aggressive subject, and in animals hallucinogens generally diminish aggression. Moreover, it is known that at least some drugs (amphetamine, morphine and morphinomimetics) release the central catecholamines from their storage granules and must therefore encourage the activity of the MFB; and morphine inhibits the synaptic release of acetylcholine, the chemical mediator of the PVS (Jhamandas, Phillis and Pinsky, 1971).[31]

Psychotic states

There have been numerous biochemical studies which have tried to demonstrate that a disturbed metabolism of the biogenic amines causes psychotic states, and in particular schizophrenia.

I am inclined to suggest a hypothesis which does not invoke any basic metabolic disturbance. The psychotic seems to me to be someone who, unable to satisfy his drives by an act of gratification on the environment, still manages to trigger the MFB as often as possible, and gains a secondary satisfaction in the imagination by mobilising memorised material. In this hypothesis, treatment ought to seek to prevent the MFB from functioning, and it must be emphasised that the anti-psychotic drugs do inhibit the catecholaminergic system (chlorpromazine, reserpine, butyrophenomes). In this kind of model, the "organic" has a very different sense from the one which people have usually attempted to give it. There are, it is true, innate disorders, but we shall be unable to define these precisely for a long time yet. And there is genetic disturbance, but this may simply be a weakness of the nervous system, a way of reacting to the environment which is due to a particular state of certain enzymatic activities in the central nervous system, an erratic metabolism either of protein synthesis within certain neurons or of the synthesis of certain neuromodulators — all disorders whose origin may be found in the genome. With these exceptions, the organic nature of the phenomenon seems to be an acquired process. But it no longer appears as a macro- or microscopic lesion of the nerve pathways or the central reserves. If one accepts that the synaptic encoding of these is carried out by protein or glycoprotein synthesis, as the basis of training and memory, then there is no point in looking for an organic lesion, even at the level of the molecule. The organic nature of mental illnesses can only reside in the establishment of preferential synaptic pathways which are metabolically encoded and result from the perpetual dynamism established from birth between the three evolving stages of the nervous system and the environmental niche. If this is the case, then the chronic aspect can only reside in the material and molecular basis of training and of the subconscious memory of conflicts between acquired automatisms and instinctive drives, but also in the fact that the rigid engramming of this conflict can no longer find expression as gratifying behaviour in harmony with the socio-cultural environment. As long as other synaptic structures can be used in response to the environment, then aggression (a reaction of the pain triggered by the PVS), repression (the exhaustion of this reaction when faced with limbic taboos) and the behavioural language of hysteria are only transient. Use of the creative imagination permits flight from unconscious conflict; but if it invades the whole nervous system, no further reference to sensory controls will be possible and madness will set in. In this case psychopharmacology is usually merely palliative, smoothing out excessive neuronal activity (whether manic or depressive).

But if one accepts this outline, the psychopharmacology of psychoses must first attempt to inhibit reinforcement, since we have agreed that reinforcement, instead of finding gratification as usual in action on the environment, finds it instead in triggering the creative associative systems of the imagination. Certainly this is how the main anti-psychotic drugs act. But by doing this they suppress the drive and direct the patient towards depression, so that treatment swings between anti-psychotic and anti-depressant drugs. Perhaps treatment might be effectively directed not at simply inhibiting the MFB and PVS, but at stimulating the PVS. In other words, it might be directed towards cholinergic stimulation, which would force the psychotic to rediscover the hard law of the environment and its travails. Aggression would probably prevail for a while. But is it inconceivable that institutional treatment might be able to recreate new synaptic relationships within the nervous system which are capable of reinforcement and the pleasure of living? Perhaps one way of approaching this programme of treatment is to facilitate the interneuronal formation of the second messenger of acetylcholine, cyclical guanosine monophosphate (cGMP). In this context it may be useful to remember that insulin increases the cerebral synthesis of cGMP (Illiano, Tell *et al.*, 1973),[32] and that electric shock is an experimental stimulant of cholinergic PVS. These are two treatments of psychosis which are not devoid of effect.

The neurotic personality or "terrain"

Perhaps this is where the classic discussion between the innate and the acquired is most often encountered. At the present time it seems impossible to deny that the appearance of neuroses is connected with their possession of a suitable genetic terrain. But the genetic combinatorial [*combinatoire*], which is linked with sexual differentiation, complicates the problem from the outset. It may be admitted that the genes simply control the enzymatic equipment, and that the congenital enzymatic "errors" so far discovered give rise to precise syndromes of behavioural deficiencies; but if at the same time no enzyme deficiency has so far been proved beyond doubt to be responsible for the neurotic "terrain" or personality, then it is tempting to think that these latter are acquired.

For the nervous system, the engramming of the environmental niche begins *in utero*, and its very response to environmental stimuli will transform its surroundings, making some of the stimuli important for some individuals but not for others. "Assimilation" and "accommodation", in the sense which Piaget[33] gave these terms, represent a perpetual feedback exchange between the surroundings and the nervous system; this exchange is enough to explain the uniqueness of each human personality, for no individual occupies the same position in space and time as another. Since we cannot make a complete list of the environmental factors capable of influencing a human nervous system (we cannot put ourselves in its place and remake its history, which in any case is significant for that individual and no other), we can simply recognise the existence of neurotic personalities, without defining how they are established. The role of hormones (ACTH and the corticoids in particular) in the transcription and translation of the genome, either directly or indirectly through cAMP and cGMP, seems to be important in protein synthesis and thus in enzymatic synthesis, on which the central nervous activity is based. The psychopathic personality seems to establish itself in response to the environment, and is perhaps thus not simply dependent on the central neuromodulators. The connections between the synthesis, storage, release and metabolism of these neuromodulators on the one hand, and the endocrine secretions on the other, in their relation to the central nervous activity, have not yet been precisely defined.

3

The level of organisation of human societies: a historical outline of domination

The human organism, the level to which we now come, cannot be conceived in isolation from its social environment. A child abandoned and deprived of human contact would never be more than a "wild child", which if found after several years and put back into the social situation would never become a human being. What we internalise in our nervous system from birth is, essentially, other people. But we internalise them in our nervous structure, whose organisation by levels of evolution I have just outlined. The innate which is given to us as humans in our DNA persists, and changes of surroundings do not change the functioning of the instinctive drives which up to the present time have organised our socio-cultural relations in the service of domination and hierarchy, as in all animal species.

It is by belonging to a social group that the individual discovers his informational opening, and this regulated system becomes a servomechanism by virtue of the information which it receives from outside itself and which regulates its behaviour.

Unfortunately social groups, families, classes, ethnic groups, nations, groups or blocs of nations enter into competition with one another in pursuit of domination and to maintain their own socio-economic structures, which give preference to certain types of hierarchy: capitalist, technocratic, bureaucratic or whatever. To put it another way, the opening to information becomes closed at a certain level of organisation; the human species as a whole is not yet the largest set, for whose survival each cell, each individual, each group of individuals is necessary, just as the set is necessary for the survival of each one of them.

It is perhaps the last level of organisation of the nervous system, the associative zones of the orbito-frontal cortex, which will enable man to conceive a socio-economic structure that can bring about the great human dream of going beyond pre-human domination and hierarchies.

I would like now to return to some of the ideas which I touched upon earlier and which need more thorough development. In my opinion, it is absolutely necessary to understand clearly the fundamental difference in the human sciences between the notion of information and that of mass and energy. No form of politico-sociological analysis (and this includes orthodox marxism and the various forms of contemporary marxism) has really made use of this difference which, in my opinion, throws a new light on the entire set of social relations.

The second important idea is that there is no hierarchy of value between the rational and the irrational. The point is not to be in favour of the one rather than of the other. The irrational only exists as a function of our ignorance of the biochemical and nervous structures which control the unconscious. The irrational is not a septic tank where we immerse the unspeakable, foul-smelling decay of our thought which we don't dare bring out in public. The unspeakable is only such as a function of moral criteria in a particular society at a particular time. We deal in value judgements: but things are content to be what they are, and the unconscious similarly is what it is without being beautiful or ugly, good or bad, useful or harmful, except as a function of the prejudices of an epoch. One could in fact add that it constitutes the deep-rooted source of our creativity, the hidden treasure of our inspired intuition and of the motivations that give rise to it.

Conversely, the rational exists only as a function of the assumptions on which it is founded, and the choice of these is generally the expression of an unconscious and therefore irrational determinism. Everything will become rational if we succeed one day in putting a little order into the source of our behaviour, if we succeed in specifying what the structures of the unconscious are and the laws of its dynamic. This is what the biological approach to behaviour is today attempting to do.

But this means to say that since we have not succeeded so far, we have only created the appearance of rationalising the irrational, of rationalising the unconscious in everything outside the science of inanimate matter, physics, and particularly in the so-called human sciences.

Thus "information", the giving of form to social structures, has always been dominated by the power instinct of the individuals who make up these social structures. The power instinct is not rationalised because it is unconscious; it is more often camouflaged by paternalistic, socialist, humanist, élitist (etc.) phraseology.

A historical outline of domination

There could not have been a very great difference between the societies of anthropoids which we are still in a position to observe and the first human hordes. The establishment of hierarchies must have come about in similar ways; the fact that what by convention is called a "human being" is in reality a possibility for handling information could not, at the outset, have played a great part in the establishment of hierarchies. However, the mere act of cutting flints was already characteristic of the human species, and meant that it was capable of giving inanimate matter a particular form; this made its action on the environment more effective. We can see that at this palaeolithic stage the distinction between thermo-dynamics and information is already clear. The same expenditure of energy can be used on any flint, to produce either a heap of smaller pieces or a flint which is cut, given form and informed in such a way that it can serve as a tool, a hunting weapon, an axe, etc. But it is probable that at this time each male individual needed to know all the technology of his time, that specialisation did not properly exist yet and that the factor controlling the cohesion of the group consisted above all in the multiplicity of means deployed in hunting rather than their diversity. On the other hand, it is agreed that a functional separation of the sexes had already appeared. The women stayed at home to look after the children, for it is difficult to imagine pregnant women and children taking part, even at a distance, in big-game hunting. Initiatory rites introducing the adolescent into the group of adult males long perpetuated this separation of the sexes and still perpetuates it today in certain primitive human groups.

Domination must at this time have been established on the basis of physical strength, skill and experience. The latter, be it noted, is already the expression of memorised information, that is to say occupational [*professionnel*] information, though the human occupation was for some centuries limited to the daily hunts which were indispensable if the provision of food was to be ensured. The preservation of meat by salting, remember, began with the neolithic period. Until then no reserve supply of meat was possible. Famine must have been the dominant preoccupation of each human group; their field of consciousness was for thousands of years entirely

63

occupied by the need to find nourishment and protection from the inanimate environment in just the same way as the other predatory species. So one may assume that, in the beginning, the invention of the tool corresponded essentially to man's desire to increase the effectiveness of his protection against the aggressiveness of his environmental niche, and to increase the satisfaction of his basic needs. This is the motivation of societies of scarcity. It is a motivation of immediate survival.

Agriculture, cattle-raising, preservation in salt and the storage of food enabled man at the beginning of the neolithic period to build up reserve supplies instead of living from one day to the next. Free time was utilised to establish increasing specialisation. This was the beginning of detail labour, which then found expression in the establishment of the artisan and brought into being an increasing interdependence between all the individuals belonging to the same community. The role of women at the beginning of this fundamental stage of human evolution is often referred to. The sedentary life of women perhaps enabled them to observe natural phenomena which the hunter did not have time to observe. The female myths of fertility seem to confirm this.

Thus there are certain human groups tied to the soil in especially fertile regions, the deltas of the great rivers in particular (since the irrigation of cultivable land remained unknown for a long time to come). The first neolithic villages with their occupationally diversified human community began to appear. But according to Lewis Mumford,[34] these first village dwellers very quickly forgot how to handle the arms which were necessary to defend them against carnivorous predators and against groups still at the palaeolithic stage, who were attracted by their accumulated wealth of food. Some of these latter groups must have taken over the direction of the first cities. Hierarchies were established, with aggression providing the leaders responsible for the security of the community, while the priests dealt in myths and ensured that it was protected by friendly gods.

The free time which man obtained in the neolithic period by building up reserves could be used for getting to know the structure of the physical environment better. Thus over the centuries he constructed physics and thermodynamics, because it was easier to look outside himself than inside. But for many years now it has been clear that physics does not constitute all of "science". We shall see later how science finds its outlet in technology, the construction and control of machines, increasing production and the industrial society known as "the consumer society". The rejection and even hatred of science shown by a large section of youth today springs without doubt from the fact that the field of action of science is reduced in their minds to inanimate matter. Only the beginnings of the sciences of living systems as yet exist; the so-called human sciences have so far been restricted to words and to the level of organisation of interhuman relations, in ignorance of the biological bases of behaviour; and no attempt at coherent synthesis has yet been really tackled, because of a lack of experimental data gathered at levels of organisation ranging from the molecule to the behaviour of the individual in the social situation. One can therefore understand the new-found fondness of the younger generation for phraseology, the irrational, the mythical, for immediate globality as against analysis followed by evolving synthesis, and for the destructuring of logic by drugs. Understandable too is their revolt against a generation which seeks to impose upon them a socio-cultural framework claiming to be founded upon a reflective consciousness, but in reality founded upon the aggression necessary to achieve domination within hierarchies they no longer understand, with criteria of submission they no longer accept, for an aim they no longer desire.

When the attraction of myths and of the irrational is not sufficient to involve them, contemporary youth "satisfies itself" with the use of known grids — either the marxist grid or the psychoanalytic one — which seems to supply a coherent answer to the fundamental questions any thinking man poses for himself today. These grids have unfortunately impeded attempts to go outside them and look further afield. They themselves are in effect at the root of certain individual and group hierarchies.

Perhaps it is time to point out to this younger generation that the rational speech which lies beyond the precise laws of matter has never expressed anything but the unconscious, i.e. our desires and our socio-cultural automatisms, that introspection simply means observing the map without ever exploring the territory, and that they themselves are simply perpetuating this inadequacy by depriving themselves of the rigid coherence of the exact sciences. Perhaps it is time to say to them that there is not only the deceptive magic of words on the one hand, and the sciences of matter on the other (actually, the technological influence of the latter is turned to full advantage by our pre-human unconscious brain, which itself guides all discourse). There is also a science which is in the process of being born. It is as urgently necessary to spread the elements of this science more widely as it was to spread the elements of arithmetic which were indispensable to commercial civilisations for the keeping of accounts. This science is the science of the living world. We shall see what this science can teach us about growth and the causes of growth.

Had we stayed at the paleolithic stage, we would never have talked about "growth". There certainly was progress in cutting flints, but sources of raw materials and the accumulation of waste would certainly not have posed fundamental problems for the survival of the species. That is to say that if these things are problems today, they spring from the fact that man the transformer of inert matter into more elaborate products has found means of using more raw materials, means of creating more waste and, in the course of this, means of transforming crude matter into a growing quantity of products of his industry. For this to come about it was necessary for him to interpose the tool between his hand and matter itself; and tools have become increasingly complex and effective, enabling him to increase the yield of his labour considerably. Supposing that people had been as numerous on this planet in the past as they are today, with the same duration of human labour but in the absence of machines, then the problem of growth would have presented itself in a quite different way.

Why, given such conditions, did man invent machines? Clearly not to reduce the amount of labour which he does, for he certainly does not work appreciably less today than in the past. One must suppose that from the outset it was essentially in order to protect himself better from the aggression of his environmental niche on the one hand, and to satisfy his basic needs better on the other. Now these are in reality fairly modest, since they are governed by the most primitive drives of the hypothalamus: hunger, thirst, mating, and the need for protection against inclement weather by means of clothing and shelter, which already are acquired socio-cultural needs. Until a quiet recent period this was the situation for the vast majority of human beings. These primitive drives are satisfied today, for a large number of human beings in the industrialised societies. This is not always the case for an even larger number in the so-called developing societies. This is partly a consequence of the appropriation of the planet's wealth by the first group at the expense of the second, who don't intervene in the process of growth or at least don't benefit from it.

Of course, the transition from the palaeolithic to the neolithic stage was due to the possibility of storing up reserves which provided an effective answer to famine, and therefore saving is on the

66

whole probably a factor of growth. But it can only be a secondary factor. To produce more, to keep for oneself a greater margin of security in the event of poverty, constitutes a motive both for individuals and for social groups which without doubt has some importance at the early stages of industrial development. But the affluent societies, for whom growth is an end in itself, are not saving societies but consumer societies. The pursuit of security cannot, therefore, be its main motivational factor. Moreover, security in our day has become a collective rather than an individual process. Greed seems to be a kind of behaviour which is in the course of disappearing at the organisational level of the individual, although it persists at that of social groups, in the accumulation of capital. But even in this latter case, it exists to satisfy the need for domination on the part of social groups and the hierarchical structures which animate them, rather than to soften the anxiety about what might happen tomorrow.

Can the pursuit of "well-being" be the cause? But first of all, what is "well-being"? It should be noted that we are talking about a relative condition. Its basis would seem to be physiological and biological. Cabanac[35] has shown that a stimulus is not pleasant or unpleasant in itself, but is felt as a function of its utility in relationships with other internal signals. For example, when an experimental subject placed in a bath is asked to indicate on a five-fold scale (very pleasant, pleasant, neutral, unpleasant, very unpleasant) the sensation he feels when he dips his hand into a bucket of water outside the bath, it has been established that he finds the cold water in the bucket very unpleasant if he is placed in a cold bath and very pleasant if he is in a very hot bath. A number of experiments of this kind have demonstrated that it is as if satiety modifies the feeling of well-being or pleasure to the point of turning into its opposite. Cabanac suggests this should be called "alliaesthesia".[36] The socio-cultural problem is identical: the result of assuaging any acquired (socio-cultural) need is dissatisfaction.

I should add that pleasure and suffering also depend on training, i.e. on the organism's increased range of oscillation around average values. Training makes it possible to tolerate heat changes of considerable magnitude (such as rapid or sustained muscular exertions) and pushes back the point at which discomfort or suffering is felt.

It is probable, therefore, that man's actions on his surroundings give rise to a regulation (as a "tendency") which results not only in the improved homeostasis of his physico-chemical characteristics

67

but also in a progressive loss of training in the variations of these characteristics, so that the tolerable range within which "well-being" is preserved narrows accordingly. This happens with air conditioning, the lift, the various means of locomotion which have replaced walking, etc. To this we must add the pleasure which comes from more rapid communications and exchanges of information, and from better health. Here we again come across the idea that the invention of the machine, which is interposed between the hand and the desired object in order to make its production easier, correspondingly diminishes the human energy necessary for its production and as a consequence pushes back the point at which this expenditure of energy becomes disagreeable. But while it increases the effectiveness of human actions on matter, it also makes man more dependent upon the machine, to the extent that his lack of adaptation to his untransformed surroundings accelerates his de-training.

But this is not where the root of the problem lies. Certainly "well-being" springs from the satisfaction of the basic needs, but I have already indicated that modern industry is not indispensable to the achievement of this satisfaction. The satisfaction of hypothalamic needs does not depend on industry or growth. If this were so, our grandparents — even in the best bourgeois society — would have been very unhappy people. This implies that well-being chiefly depends on training. Thus it becomes a socio-cultural concept. If you could ask a palaeolithic man what he had most "need" of, he would doubtless reply: "A bear for every meal and a bit of fire to cook it over." He would not have asked for a new motorway. In reality the concept of well-being is intimately linked to the concept of needs. But if the basic needs are already assured, needs are necessarily linked to the knowledge of what one can wish for. All advertising is based on this necessity of making something known in order to arouse a need. One cannot wish for what one doesn't know. On the other hand, one can wish for what another person possesses and what one does not possess oneself, particularly if possession of the object allows one a place in a hierarchical order and plays a part in establishing domination. Possession produces two effects. One is the satisfaction of a need which is not basic, a desire which has been learned, and the achievement of a well-being created by society. The other is a means towards obtaining this well-being through domination, the bio-sociological meaning of which we shall shortly examine.

So the problem consists in understanding how the myth (which has so far kept its motives hidden) of growth for growth's sake and not only for the satisfaction of the basic needs came to be established, how it has come to be taken as a basis for social behaviour in the industrialised countries, and how it can today be championed as an end in itself — indeed, as the finality of the human species, clothed in such emotional and mystical concepts as "happiness", "progress", man's domination over cruel Mother Nature, the genius of the white race or some particular ideological system. All this is defended by a perfectly rational discourse based on *a priori* arguments and value judgements, such as "social advance" (always seen as a good thing in itself), free competition (for in a world said to be "free", competition must necessarily also be free), international competition, work as a virtue, wars (which supply a daily ration of bravery and heroes), the defence of tradition and of the currency, etc.

Some people seek analogies with the growth of the foetus and the child, and are pleased to be able to point out that these are programmed in the DNA of the fertilised egg. But when they get to the socio-economic level, they say there is no longer a programme, because the activity of complex or hyper-complex systems has unforeseeable consequences. Indeed! They may be unforeseeable to rational discourse, but the mere fact that socio-economic processes are not aleatory events enables us to see that they are programmed. And they are programmed chiefly by the unconscious bases of human behaviour which rational discourse has not so far brought into the reckoning.

The machine, an outcome of the technical development which is itself the result of knowledge of the laws of physics, is one of the "means" which enable us to secure the growth of consumer objects and collective goods. But the machine is not the "cause" of growth. The cause can only be man's *behaviour,* which pushes him to produce more and more. The machine, which with the arrival of electronics and information science leads to the automation of labour, has no desires, no needs and no socio-cultural automatisms. It simply obeys the desires, needs and socio-cultural automatisms of man, who builds it. On the other hand, detail labour springs from the generalised usage of the machine, and by obscuring the meaning of the work of the individual, it is certainly responsible for the fact that man searches for some kind of compensation through the possession of more and more objects. The machine may therefore be an indirect "cause" of growth.

In the social situation, domination is necessary in order to satisfy instinctive drives which, in the case of the animal, are primarily eating, drinking and copulating. Domination is therefore necessary for the achievement of well-being, and animal hierarchies thus seem to be based on an aggression which disappears when domination and hence the instinctive drive have been satisfied.

Delgado[37] has shown, by stimulation of the amygdala, that aggression is linked to a state of suffering and displeasure, and that it is only exerted in this case on the subordinate, not on the dominant. The grooming (the search for fleas) performed by the dominant animal is the general act of obtaining submission from and a simulation of the sexual act upon an animal of the same sex, the primitive gesture of domination. It is worth noting the curious fact that man uses certain gestures of the forearm — they are little used, I admit, by the aristocracy — or certain insults, which likewise are not the everyday language of academics, to indicate to his fellow human-beings a type of behaviour which actually is performed by chimpanzees.

So one may submit that things are fundamentally the same for man: but man uses language, which transmits information from one generation to the next. The language of gestures fades away with the performer, but written language enables man to institutionalise the rules of domination. This is how moral and ethical rules, prejudices, value judgements and the laws which govern the behaviour of individuals in a society in a particular epoch become institutionalised. It is not, of course, the dominated who impose their laws on the dominant. The "culture" of an epoch therefore represents the rules to which an individual must submit in that epoch in order to ascend the hierarchies and achieve domination. Without this domination there is no hope of reward, no hope of pleasure or biological "well-being". The point is no longer to secure the basic needs, which the so-called developed societies are now more or less in a position to secure for the majority of individuals, but to secure the individual's "freedom of action", which is a function of his power in the group to which he belongs. When people speak of the "full flowering" of the individual, are they also aware that such a utopia is unrealisable within the framework of any kind of hierarchy?

Why, in our highly hierarchical societies, has there been such an explosion of so-called "psychosomatic" diseases? They are simply the bodily expression of conflicts in the central nervous system between instinctive drives and socio-cultural prohibitions; these conflicts cannot be resolved by effective and "satisfying"

70

action on the surroundings, because the dominant have institution-alised the rules of domination.

It is these rules that are the fundamental factor in the emergence of industrial societies and of the myth of growth, as I shall now try to demonstrate.

Domination was for a long time based on criteria such as that of physical strength. The selection of leaders was based on certain warlike qualities, in epochs when the social group, acting as a predator against other social groups, needed a leader with whose personality each individual could identify. Under these conditions, there was little cause for growth to proceed at a gallop. But the inadequacy of this hierarchical system of domination and its privileges was exposed when the development of trade concentrated wealth in the hands of what one may call the bourgeoisie. This class possessed none of the attributes of the aristocratic class which preceded it, but possessed a greater power: the power of capital and private ownership of the means of production. It became intolerable and "displeasing" for this new class to remain subject to domination by a class which was no longer indispensable to the social equilibrium.

The French Revolution expressed the aggression of this class, subject to a domination which had become intolerable because it was useless. It was intolerable also because it limited the gratification of those who could not become members of this class. We know that aggression is one of the most primitive means of resolving the impossibility of action which stems from the conflict between the drive to dominate and socio-cultural prohibitions. The immortal principles of 1789 — the rights of man and of the citizen — institutionalised the rules of the new domination, the rules that had to be respected in order to become a bourgeois. The most insignificant Frenchman can, we are told, hope to become President of the Republic one day. But they forget to add: if he obeys the rules of the game, the value judgements institutionalised by the bourgeoisie, and in particular private ownership of the means of production. From 1789 onwards, all the hierarchies — bureaucratic, military, academic or whatever — have existed only as a function of the central hierarchy, that of money. Money is what affords the greatest power, and in particular the only power which can be transmitted by inheritance. Of course, the socio-cultural automatisms of the bourgeoisie, the innumerable recognition signs which constitute "good breeding", are transmitted more easily to

71

someone who is born into a bourgeois environment. But these signs can be learned, even if one comes from a different environment. The only thing that can be directly transmitted and allows domination by inheritance, without effort, is property. From then on the whole social edifice is subject to it. All the hierarchies bow down before it, since they are now its mere tools. In order to add to one's wealth in such a system, and thus to protect one's power and domination, it is necessary to sell. It is necessary to appropriate the labour of the dominated and, by means of surplus value, to expand the capital which secures domination. In order to expand this capital, the dominated must produce wealth and this wealth must be "consumable". In fact today, those who are dominant by means of capital are not sufficiently numerous to consume enough wealth on their own, and so it is necessary for the whole of the population to consume in such a way that profit increases and that power — as a function of profit — increases too.

This, in my view, is what lies at the root of the problem of growth. One can in fact see how there is an instinctive biological satisfaction which in the social milieu cannot be achieved without domination, and that with this notion one passes from the individual to the social class which institutionalises the rules of domination and the hierarchies. One can see how, from the moment domination demands the possession of capital, this latter can only be obtained by the appropriation of surplus value, but that if it is to grow there must also be a growing production of consumer goods and the participation of the producers themselves in this increasing consumption. One can see why under such circumstances the desires of the dominated, of the masses, are concentrated on satisfaction by means of goods and the ownership of objects, and why there is an attempt to assuage their aggression (which springs from their complete lack of power) with the trivialities that pass for consumer goods. This type of contemporary society has gradually and completely concealed its initial motivation — i.e. the pursuit of individual pleasure by means of domination — and is entirely alienated from the means for obtaining it, i.e. production for production's sake carried to such a point that it becomes a finality in itself and the only way of satisfying needs. But the only need which is essential, which is not satisfied in a general way, is not consumption but power. Only power makes it possible to satisfy ceaselessly accumulating needs, because only power can supply the knowledge of how to satisfy them.

The evolution of contemporary societies offers a great deal of confirmation for this view. So far it has been the possession of capital that has guaranteed domination. But precisely because the causal importance of this factor, as well as the means of reaching its goal (i.e. production), have been concealed, a new social class has been tending to take power from the capitalists, namely the various kinds of technocrat. They are indispensable to commodity production, and have recently become conscious of their power; they are less inclined to accept the domination of the "capitalist", who seems to be becoming useless. In reality the big capitalist families are diminishing in number, and it is the *management* of capital — increasingly internationalised and spread among more numerous owners — rather than *ownership* which has become the means of achieving domination. Here is proof that the pursuit of domination and "power" is indeed the fundamental motivation of man. When it is no longer necessary to secure the possession of capital, and when "material" well-being is seen to be sufficient, then what our dominant contemporary seeks to appropriate is the management of capital, the decisions about how it is to be used (investments), and the knowledge of production techniques for the objects which make the growth of capital possible.

This new means of establishing domination obviously changes rather profoundly both the rules for obtaining domination and the hierarchical systems. But it in no way alters the pivot on which the whole issue turns: the production of material goods. If this ceased to be the finality of all human activity for our contemporary civilisation, all the existing hierarchies would vanish. Even the politicians, who certainly still cherish the illusion of having a certain power, are mere puppets in the hands of a faceless monster: production for production's sake, or, as it may also be termed, capital. But this is an increasingly impersonal instrument for securing domination.

All the same, one must recognise that even today the possession of capital allows the capitalist to fit himself out with technocrats if he has no professional qualifications himself.

It appears difficult to relate so many secondary processes to a single biological axis, i.e. the pursuit of domination: the improvement in the living conditions of the dominated, their increasing participation in the material product of their labour, the gradual depersonalisation of capital, the increasing grip of blind technocracy and bureaucracy. It is especially difficult because this common axis on which behaviour hinges is never actually invoked. Consciously or unconsciously it is camouflaged beneath a deceitful phraseology: the full flowering of the human personality, profit-sharing (but not power-sharing), freedom, equality of opportunity (i.e. the opportunity to become a bourgeois, to participate in the ownership of capital — just when the pleasure and the reward are tending to pass from those who possess to those who make decisions).

But to return to a point I made in *L'homme imaginant*, this decision-making power is itself questionable if it is no more than the power to decide on a better means of making a few more commodities and a bit more profit, so as to increase the investments which make it possible to increase the production of commodities and therefore of profit, and so on ad infinitum.

The drama of this situation stems from the fact that this kind of human behaviour, which is animated by what I shall no longer hesitate to call the "myth" of production, not only governs relationships between individuals but also those between social groups. The strongest groups, which are the most productive of consumer goods, engulf the weaker ones; and at a higher level of organisation, nations and even blocs of nations compete over their productivity.

It is well known that for some time now the ecologists have been drawing attention to man's accelerating destruction of the biosphere. I think we can summarise the whole ecological problem in its infinite complexity as follows. The biosphere is a consequence of the transformation of solar photon energy by living systems. At the level of energy, the whole evolution of species springs from the transformation of this photon energy into chemical energy, in increasingly complex systems which culminate in man. Man cannot simply open his mouth and feed himself on solar photons. To feed himself he needs all the life-forms, from photosynthesising plants and marine plankton, through the bacteria which recycle organic matter, to numerous insects, birds and mammals. The relationships among these multifarious life-forms constitute the eco-systems, whose frail balances are only now beginning to be grasped. The benefit of increased industrial production is often only a short-term one, but the ensuing drama will be of long duration if this production clumsily destroys the balance and disturbs a stage in something which is indispensable to man: the transformation of solar photon energy into the chemical energy which he uses to feed himself.

Attention has also been drawn to various other consequences of growth: the exhaustion of energy resources, the accelerating accumulation of waste which cannot be recycled within the great cycles through which matter passes.

This is what happens as a result of the pursuit of domination and the myth of consumer production. They demand a massive concentration of human beings in the heart of each modern megalopolis. For the profit of the dominant (and it is the pursuit of domination which is the profit motivation), there is pollution of humanity's collective goods: air, water, the built environment and the environment of sound. Not only the pursuit of domination but interhuman relationships in all their forms have now reached the point where they constitute a real threat to the human species in its entirety.

4

Hierarchies of value and hierarchies of function

I have insisted at length on the difference between what I have called hierarchy of value and the hierarchy of function or complexity. The organisation of living systems presents us with hierarchies of function and complexity, but not of value. It is man who speaks of some organs which are noble and others which are not. An organism can live without a cerebral cortex, which is held to be a noble organ, but not without a liver, an organ which one tends to think of as plebeian. The structures of an organism are content to exist and to work properly. They know nothing of good and bad, beautiful and ugly — value judgements that only have a value for the survival of particular social groups in a particular epoch, and not generally for the individual or for the species. The nervous system is indeed situated at the pinnacle of the hierarchies of complexity, but it does not "issue orders": it is content to be a link between the environment and the reaction of an organism to this environment, between the "stimulus" and the "response". Nothing is born in it, *ex nihilo*. Everything comes to it from outside itself, except for its genetic structure (which it does not have the job of assembling). The pursuit of biological equilibrium, satisfaction and reward, of what one may call pleasure, always remains the fundamental purpose of an organism, and the organism secures this by its action on the environment. In reality, therefore, it secures its own survival in an environment whose physico-chemical characteristics are less well regulated than its own. What gives the impression that an interior source of decision-making exists is in fact the memory process. But all we need to remember here is that a new-born baby without any experience is unable to decide anything. All it can do is to suffer direct or indirect aggression from its environment, and to record in its brand-new memory the state of well-being which comes from satisfying a need and making good what it lacks, or the state of "ill-being" which comes from the failure to satisfy, from the perpetuation of internal disequilibrium. Along with its memory of gratifying or punishable acts, of satisfied drives and parental prohibitions and later of socio-cultural prohibitions, an internal nervous structure gradually becomes established as a result of the coding laid down in the neural pathways utilised in the course of these multiple experiences. But because this activity of the nervous system remains unconscious (only the logical discourse which gradually overlays it becomes partly conscious), one gets the impression that there is an autonomous "order" given to the actions of the individual by his nervous system, as if there were a little internal god who decides "freely" about our behaviour and our choices, when in fact these are only the unconscious result of our drives and training. Even when the orbito-frontal brain, our properly human brain, allows us to escape from these automatisms by providing our action with a new solution in the form of an original "pattern" of behaviour, this latter is "chosen" only as a function of the basic motivation, which is the pursuit of pleasure and therefore of domination, and which is modified by training in the socio-cultural rules that must be followed in order to obtain rewards. These innumerable judgements which govern our actions and dictate our attitudes towards the actions of others are "value judgements", because in this case it is necessary for us to have a "scale of values" in order to classify our actions and those of others in such risky categories as good and bad, beautiful and ugly, just and unjust, and so on. So things and beings are content to be: it is we who judge whether they are good or bad.

They cannot be judged in relation to absolute criteria, which do not exist except perhaps as a limiting factor to the specific set, i.e. the species, inasmuch as by belonging to it we judge it useful to protect its existence and evolution. They can only be judged in relation to our mental structure. We have just seen what this mental structure consists of: the pursuit of pleasure, the avoidance of displeasure, obedience to the automatisms which our particular social group has introduced into our memory by means of training in the maintenance of its structure. These value judgements, therefore, are only what they are because they express the rules for maintaining a closed structure, a sub-set which is not integrated in the human set. How can we account otherwise for the fact that one may often be against abortion and for war at the same time (a just war of course)?

In a system — such as the individual — which is closed at the level of the information-structure, all the organs, all the systems, all the cells and all the molecules unite to maintain the structure. The nervous system only expresses their common will not to suffer, by securing the organism's motor functions in relation to the environment, and by enabling the organism to act on the environment in such a way as to conserve its biological equilibrium. The nervous system is simply the executant of the decisions opposing entropy which are made by the organism as a whole.

This simple idea, which is so difficult for many people to accept, takes us far away from free will. It shows us that living systems in the biosphere have succeeded in bringing about *self-managed* structures; it is a matter for astonishment that while the blind determinism of biological evolution has succeeded in bringing about such systems, man in his societies has not yet been able to do so. Let us try to understand why.

In a living organism, each cell, each organ, each system issues no orders to anything. It is content to inform and be informed. There are no hierarchies of power, only of organisation.

The term "hierarchy" in this context should in fact be dropped, for it is difficult to rid it of all value judgement. It should be replaced by "levels of organisation", that is to say levels of complexity: the molecular level (comparable to the level of the individual), the cellular level (comparable to that of the social group), the level of organs (comparable to the level of human sets which assume a certain social function), the level of systems (nations), the level of the whole organism (the species). Each level is in a position, not to exert "power" over the other, but to associate with it so that the whole set functions harmoniously in relation to the environment. But for each level of organisation to be functionally integrated into the set, it must be informed of the finality of that set and what is more, it must be able to play a part in the choice of this finality. When we speak of choice, we are not speaking about the expression of free will.

For an organism, the point is to undertake specific action in response to a given stimulus, an action capable of maintaining its homeostatic equilibrium in relation to the environment; what is important is its organic structure, whose maintenance is expressed by pleasure and reward. For a social organism, accordingly, the point is to broadcast information to all the members that constitute it, whatever their functions may be.

But what I mean by "informational society"[38] is not the specialised information which enables the individual to really transform inanimate matter, nor the information which is supplied by manual or conceptual training, but a far more vast scale of information which concerns the importance of the individual as an individual within the human collectivity. Specialised information can only provide the individual with a specialised power within a hierarchy: it prohibits him from taking part in "political" power. The latter kind of information, on the other hand, enables him to enrol in a functional class and to take part in the decisions of the organic whole. Power is knowledge.

The "political level" is that which concerns the importance of the labour of each person integrated into the set, and the finality of this set in the more complex sets encompassing it. At this political level, a highly specialised engineer often has no more knowledge than a skilled worker, although the kinds of knowledge they have are different because they are dictated by value judgements and by the prejudices necessary for the maintenance of hierarchical domination. Unfortunately, specialised information thus appears to be the source of political power, for it is the chief source of the hierarchies (even if its specialisation makes it unable to enlighten political power). But we shall see later on that this is mere appearance.

In my organism, my big toe cannot replace the "functions" looked after by my liver, and my spleen cannot do the work of my heart: that much is certain. Is my liver "better" than my heart or my spleen, or does it give them orders? It is simply responsible for a different function because of its occupational specialisation. To put it a better way, the nucleus of every cell in every organ contains the whole of the genome pool, which means that it could give birth to a cell to replace any function whatsoever. It does not do so because its functional potentialities are "repressed" by certain molecules which forbid it from performing any other function than the one passed down to the organ in which the cell containing this nucleus finds itself situated. But it has proved possible, with the frog for example, to use microscopic dissection to remove the nucleus of a fertilised egg before cellular division has taken place and to replace it with the nucleus of any frog cell. In fact the nucleus of a neurone or of an epithelial cell from the intestine is used. The absence of inhibitors in the fertilised egg, whose nucleus has thus been replaced by one from a random cell, enables the latter to give birth to a frog which is strictly identical with the one from which the grafted nucleus came. This means that if the same procedure were carried out on a human egg, if it were able to continue growing after being implanted in the uterus, and if the grafted nucleus had been taken from a cell in (for example) Einstein's body, then the product would be a human being who from the genetic point of

79

view would be strictly identical with Einstein. It is striking to think that because the evolution of this individual would be realised in another time and place, in a space-time necessarily different from that in which Einstein lived, this human being would still be completely different from Einstein. Here we touch on the problem of the innate and the acquired, which is not our immediate concern. The point of this example is simply to show that *functional specialisation* in a living organism, which is equivalent to *occupational specialisation* in a social organism, is not accompanied by any particular *value* and furthermore that it in no way secures the ability to act separately from the organic set. The latter has to keep the functional specialisations constantly informed of the requirements for its survival as a set; at the same time, conversely, the functional specialisations must keep the set of the organism informed of what is necessary to ensure their own functions. This idea of a dual circulation of information from cell to organism and from organism to cell is fundamental. Specifically, the dual circulation is brought about by means of the nervous and endocrine systems, and is a result of the fact that the entire set of cells in the organism swims in one and the same liquid matrix, the *internal medium*, which conveys a large part of all this information. The internal medium can also be considered (it is its chemical constitution which allows us to do so) as that part of the primitive ocean in which the first unicellular beings appeared, which they imprisoned by uniting into pluricellular organisms and which they took with them when they progressed from living in the water to living in the air.

In this type of self-managing society which every pluricellular organism constitutes, one can therefore see a clear distinction between *specialised information* and *generalised information*. The specialised information of a cell (an organ, a system) is only a minute fraction of the set of global genetic information contained by its nucleus, and originates from the place assigned to it by ontogenetic evolution. Its generalised information, on the other hand, originates from all the other cells in the organism and keeps it constantly up-to-date on the state of well-being or suffering in the entire set of cells, so that it can adapt to its own specialised function to the pursuit of the overall equilibrium which has been lost or to its maintenance in a given environment. The point is not for it to extend its own particular functional (I was going to say occupational) knowledge, but to extend the knowledge which stems from the functioning of the organic set. There is no superior hierarchy to give it orders: it is constantly informed of what it should be doing, according to its place and its role, so that it can co-operate in the smooth functioning of the entire set. At the same time, conversely, it keeps this set constantly informed of its basic needs, the needs which are necessary to ensure that it functions correctly.

It should not be assumed that the relationship between the living organism and the social organism is merely one of analogy. The social organism is itself a living organism at a higher level of organisation and, this being so, the living organism also constitutes a genuine "model". Furthermore, the nature of the models is the same, since they belong to the same domain. An organism constitutes a cellular "society" in which the element is the cell, just as in a human "society" the element is the individual. Since the cellular society shows us the harmonious functioning of a non-mechanical social model, it would be interesting to find out what the dynamic principles of this harmony are, in order to see if they can be used with regard to human societies. The point is not simply, as in biological experiments, to transpose what is discovered at one level of organisation to the level of organisation above it (see page 28), but chiefly to understand in what ways and for what reasons the level above, the social level, does not at the moment behave in the same way as the biological level.

What are the new attributes which appear at the level of organisation of societies? Are they indispensable, are they linked to the very structures of the biological level? I have already mentioned (see page 33) the probable symbiosis that must have occurred at a stage far back in biological evolution between the new aerobic forms such as bacteria, and the old anaerobic forms. The aerobic forms must have given birth to the mitochondria. These latter have not conserved all their original properties; they have adapted to their new biological situation, but this adaptation has of course taken millenia to happen. Is the individual today still at the stage of the primitive mitochondria? Must he go through a profound transformation in order to be integrated into more effective human societies, and how will this take place? What must he do to bring about the planetary organism which the human species might become?

Let me go even further. I willingly grant that for the present I should be debarred from making any attempt to find a structural analogy. On the other hand, what biology has made me understand and what, I maintain, can no longer be overlooked by the human sciences, is the notion of information as distinct from thermodynamics, and the notion of open and closed systems from

81

both the thermodynamic and the informational point of view. Here we find structural laws which no discipline can be allowed to ignore any longer.

5

Thermodynamics and information in sociology

I have suggested that modern societies, whatever their political denomination, should be called "thermodynamic societies", to contrast them with what might be called "informational society". On page 20 I used the example of a telegram addressed by a sender in Paris to a recipient in another town in order to explain the difference between information and thermodynamics. If the letters which make up the telegram's message are put in a hat, shaken up and presented as they happen to be drawn out of the hat, there is little chance that they will constitute *information* for the recipient. But at the thermodynamic level, that is to say at the level of the *energy* needed to transmit a given number of letters by telegram, there is no difference between the informative message which is given form and is a conveyance of information, and a random sequence of letters devoid of any "information". Information demands conveyance in the form of energy and matter: but it is only information, it is not mass or energy. It represents the "structure", that is to say the set of relationships which exist between the elements of a set (these elements *can* be mass or energy). When someone transforms inanimate matter into a product of his industry, then not only is there an expenditure of the energy which has to supply the labour power he represents, there is alongside this the information which he contributes; he contributes it by establishing new relations between the elements of the inanimate matter which he handles and out of which he makes the tools, machines and consumer products of his labour, and by giving them a "form". Here, of course, we come across a difficulty which marxism often encounters: that of bringing "intellectual" and "manual" workers together. The difficulty lies in the variable relation between information and thermodynamics which exists in human labour, and in the difficulty which there used to be in measuring information (for information theory only dates from Shannon's formula of 1948, when the energy released could finally be measured with precision). From the point of view of thermodynamics, the labour of a donkey going round and round a pump to raise water from a well, for example, is performed and measured by the quantity of oats necessary for his nourishment, which is balanced to ensure that he does not grow thin (taking into account the work he does). But the genius of the person who first dreamed up the system which uses the donkey's energy by means of the pump was not measured. Moreover, the recognition such genius receives from a human group will always depend on the value judgements, needs and prejudices of an epoch; and it will find its most complex expression in the work of art, or in basic (not applied) scientific creation.

As Marx quite rightly said, man is an animal which makes tools; but the important thing in the making of tools is not so much the labour, the mechanical strength necessary, but the information which man supplies to the inanimate matter. And this has been the case from the first ages of man. The whole evolution of the human species has taken place by means of the constant increase of information which has been granted by the transmission of acquired experience through language.

However, what information does, its role, its necessity, its meaning: these things have only recently been understood. This is because for centuries man handled mass much more than energy. In fact, in order to transform inanimate matter (mass), he at first used his own muscular strength as energy. A friend of mine who is a teacher of history has pointed out to me that because the Romans had plenty of slaves, they gave up using the harvester invented by the Treviri, a tribe of Gauls who lived in the Belgian part of Lorraine and who, not having any slaves, had found themselves confronted with a labour shortage. It was undoubtedly this shortage that pushed them into inventing the harvester. This is an example of what we saw above: information is only used as a function of the needs, prejudices and value judgements of a particular epoch.

However, man soon resorted to the muscular strength of domestic animals such as the horse and the ox. For centuries, he used information to handle mass rather than to master energy. It must be said in his defence, however, that the discovery of agriculture and cattle-raising was a considerable informational advance which, by enabling him (albeit in a very indirect way) to use solar photon energy to his advantage, allowed him to progress from the palaeolithic into the neolithic age in one leap. We can in fact see here the first utilisation of *energy;* but it was an empirical utilisation. The slowness of this development is probably explained by the fact that energy is more *abstract* than mass, space and time; the latter were

84

easier to measure, and gave rise to the first rudiments of mathematics. At the outset, mathematics was essentially the language of primitive physics. It was by measuring the fields covered each year by the Nile floods that Thales provided geometry with its beginnings.

In the middle of the nineteenth century, man became conscious of the possible importance for him of mastering energy under its various forms. This led him to realise, through his knowledge of the atom, that mass and energy were simply two different states of one and the same thing, as Einstein's equation was to state. With this awareness of the importance of energy, man made the breakthrough into industrial civilisation.

But he made it without grasping the significance of information; he has thus prolonged the era of thermodynamics right into our own times, controlling energy in the same way as he had always controlled mass and trying to solve his psychological and social problems with this encumbrance.

Social man has been judged by his commodity productiveness for a long time. We have believed that what characterised him was the transformation, by hand and later with the aid of tools, of matter into manufactured products and later into industrial products. His labour power has been bought and sold strictly at the thermodynamic level. The problem of labour value has become more difficult to solve since the machine has supplied the expenditure of energy which for a long time had been man's. The nervous exhaustion that results from the work rhythms currently imposed on the worker is difficult to evaluate qualitatively, and it is "qualifications" that pay. Basically it is the quantity and quality of information stored in the worker's nervous system by training that are paid for. Increasingly it is the machine that performs the "energy" part of the production process; man is increasingly kept for the informational part. It is even possible to imagine a future when, thanks to automation, all labour will be the province of machines, and information will be the province of man, whether it concerns the construction and control of machines or the fundamental scientific progress on which their constant improvement will be based. Where this kind of development remains within the hierarchical framework that exists at present, "power" and domination will probably always be distributed according to the quantity of information *supplied* by the individual through his creative imagination, or *reconstituted* by him as a result of his training. From the hierarchical point of view, it is more valuable to graduate from a university than from a polytechnic, to be a "scientist" rather than an "engineer". Even supposing such a social structure could distribute industrial production on an equitable basis, even supposing that it could do so without hierarchical distinction, the distribution of power and domination would still not be changed. We should therefore remind ourselves yet again that an organism contains a hierachy of functions and of complexity, but not of value.

In what I have just said I made a distinction between the information *supplied* by an individual through his creative imagination and that which is *handed back* by him after his training. I shall return to this idea at the end of this book when the question of creativity arises. In the meantime it should be noted that no personal element is involved in the individual's restitution of the information supplied to him by his training (a training which is supervised, at all levels of abstraction from the workshop to the most important institutes of learning, by means of exams and competition). The human brain simply utilises its power of abstraction to a greater or lesser extent. In general it depends on the environmental niche in which the individual has grown up, in other words on the socio-cultural and family environment which make the individual more or less skilful, more or less motivated, able to handle information in a more or less abstract manner. Generally speaking, the degree of abstraction that he attains in the handling of information is inversely proportional to his thermodynamic participation in the labour of the social set. But even this abstract handling of information is increasingly coming within the reach of computers, to such an extent that the anxiety about whether man will one day be dominated by robots is a commonplace.

Let us note in passing that it was man who invented computers. In so doing, he has not simply handed back the information supplied to him by his training, he has supplied new information. He has organised in an original manner the elements which preceding generations put at his disposal. He has used *imagination*. But in all probability this imagination does not exceed the possibilities or the performance of the machine; and yet the machine cannot be endowed with human behaviour. What the machine in fact lacks is motivation. It can be endowed with memory and a most effective associative system. But in man, the gathering of memorised facts does not take place accidentally. There is a motivation, which is to protect his structure — we have called this the pursuit of biological equilibrium and pleasure. The day man is able to create machines whose finality is self-preservation and self-reproduction as machines, whose finality does not lie outside them, whose gathering of information is not dictated by man but by the machine's own "desire", then at that point certainly man will have produced a model of his own behaviour. Furthermore, it must have a level of organisation which will already be extremely high, i.e. not a molecular level. The best means of building a human model starting from the molecular level will remain, for a long time, making love: that is, until we can control fertilisation by developing the foetus artificially in a test tube, on the basis of an artificial genetic set.

I have mentioned the famous Shannon formula on information. However, if we refer back to chapter one, we shall find that the way in which we have used the concept of information, dictated as it is by our observation of living systems and our attempts to understand the way in which they are organised, by no means carries the same meaning as the concept of information employed in Shannon's formula. This is because Shannon's theory is a theory of communications enabling the engineer to avoid the loss, through noise, of a part of the information transmitted in a message, whereas what the biologist is interested in is the actual structure of the message. The biologist is interested in the *dynamic structure* of the signifier; the information scientist is interested in the *protection* of the signifier with specific reference to the phenomenon of interference. If we revert to the example of the telegram, the biologist is interested in the relations between the letters, in the composition of the phonemes and the monemes of living systems, in their articulations and structures. He does not have to look for a transmitter and receiver, for there is no message to transmit or to protect from interference. There is only a form to observe. Only the transcription or translation of the genetic message presents some kind of analogy with the information of the information scientists.

At least, this is how it is until we get to man. Then the semantic content of the term "information" starts to shift. In fact the human nervous system, which endows the individual with his behaviour in society, is well organised, in a manner appropriate to the species. In this sense, of course, he is genetically programmed and is the bearer of information which summarises his structure and subtends his functions. But in addition to this, his structure is programmed in such a way that it is self-transforming — self-programming, as it were — on contact with its animate and inanimate environment. His information grows with experience. His neuronal coding is enriched, and his associative processes give him scope for the imaginary,

87

enabling him to enlarge his memorised structures with products of his imagination. This nervous system can thus be in a constant process of reshaping and structural enrichment, on condition that it is not enclosed in its own automatisms, i.e. that from the standpoint of its circulating information it remains an open system. This re-forming of the nervous system ensues from the "information" which reaches it from the outside world and which is enriched in turn by new structures which come from the functioning of the associative systems. We are now dealing, therefore, with *meaningful* information, not with genetic information or the structure-linked programme in their narrow sense. But it is not as simple as that. This information is only meaningful because our sense organs can perceive it (what is ultrasonic we do not hear), and because it is structured by the nervous system which introduces informational order into the world around it. Without this order, its action on the environment would be ineffective and incapable of securing survival, biological equilibrium or pleasure.

Why is this information meaningful? It is not meaningful merely because previous experience and the memory of it have given a meaning to survival, biological equilibrium etc. The signified is added to the meaningful "signifier". It is meaningful because the nervous system has established relationships between the variations in energy occurring in the environment and the biological, physico-chemical structure of the organism. The nervous system of an animal is often able to pick up variations in energy just as well as the nervous system of the human being. Like the human being, it can memorise them. It can establish certain spatial and temporal relationships between them and the realisation of reward or punishment. In other words, it can build up conditioned reflexes. But what it cannot do is, on the basis of experience, to imagine material, positional or conceptual structures which would allow it a projection on the environment whereby these imaginary structures could be realised, and whereby its action would then put it in a position to discover new structures. The animal is hardly capable of making working hypotheses or carrying out control experiments. And since in particular it lacks linguistic and mathematical abstraction, the animal is incapable of conceptualisation or generalisation in its handling of structures imposed by the environment.

We can see, therefore, where the semantic shift in the use of the term "information" has its basis. From biological information linked to the structure of living forms, we proceed to the structures which man imposes on the world around him through his action on

88

the environment, which is based on the imagination. The human organism thus becomes a source of information, a generator of structures. But in order to do this, it has to abstract information from the environment by discovering the laws which govern the apparent chaos of the world. While man's nervous system feeds on the information and structures which come from its contact with its surroundings, it is equally the case that by means of its action on those surroundings it controls their effect. Out of the set of relationships that constitute the structure of the world around him and represent what we may call "the real", he singles out only one subset; but he enriches this with the thread of his own history by means of language, which enables acquired experience to be transmitted down through the generations. And this sub-set enables him to act, in other words to transform and give form to the world in which he lives. The transmission of acquired experience by means of language brings us once more to the notion of information as message and the example of the telecommunications engineer.

Therefore when we proceed from individual structures to social structures, the mere existence of the memory of the nervous system and the abstraction of language means that we are passing from an invariant genetic "information-structure" to a relational and cybernetic kind of information which is retroactive on the environment. It would seem logical to map out at this new level of organisation an *information-structure* of societies comparable to that of organisms, and a system of *circulating information* comparable to the nerve and endocrine messages which enable the organic set to realise its finality. Of course we know that such a society does not yet exist. Occupational information is apparently the only kind of information that is disseminated, since it enables new hierarchies and dominations to be set up and rules out the cohesion of the functional social groups. It allows only the production of commodities, for it is commodity production which is the intermediary in the establishment of domination.

In other words, ever since his origins man has found himself in a perpetual feedback with his environment. Because of his nervous system and in particular because of his orbito-frontal lobes, which are capable of imagining new structures out of facts memorised in the course of his contact with this environment, he has been able not only to discover gradually the "information-structure" contained in the environment, but also to add information to it. As he has made better use of energy, he has structured inanimate matter, mass, more effectively. This two-way activity, from the environment towards his nervous system and from his nervous system towards the environment, can be considered almost exclusively as the result of the nervous system's ability to process information. In fact it is his growing knowledge of the structures of the world around him that has enabled him in turn to structure this world better, for the purposes of his survival. At least until recently. But he continues to be ignorant of the functional structure of his nervous system, and this prevents him from carrying out any effective action on himself and on his relations with his contemporaries. And so he has used specialised information, which produces commodities and transforms the biosphere, in order to perpetuate the hierarchical social structures which he derived from the species preceding him. He has yet to discover generalised "circulating" information and to broadcast this knowledge to the members of his collectivities. Perhaps this is how man will find his way to the level of organisation that lies above competitive and aggressive social groups, creating in practice the planetary organism of a human society (see figure 8).

Let us use the distinction between information-structure and circulating information once again, in order to bring up another piece of evidence that seems to be of fundamental importance in sociology. First of all it should be noted that the information-structure, which comes from the genetic programme, "circulates" too, but on a very different time-scale, since it is transmitted from generation to generation, enriched by the genetic combinatorial which sexuality affords.

We have seen that the information-structure of the individual, however, is relatively stable, and it follows therefore that it is relatively "closed". Its opening on to its surroundings transforms it only in a very limited way, by means of what memory and experience contribute to it. The "specific" information-structure on the other hand, namely that which makes it possible to say that an individual belongs to a certain species, remains unchanged. Thus the structure is open from one level of organisation to another, from the molecule to the organic whole. *But at the limits of the individual it closes.* While the individual is indeed open to his surroundings, whose changes he records and on which he acts, and while "circulating information" between him and his surroundings certainly exists, he is nevertheless virtually closed from the standpoint of "information-structure". Now as long as a structure is closed we can be sure that it can only continue to exist if its finality, its *raison d'être* one might say, is the maintenance of its

90

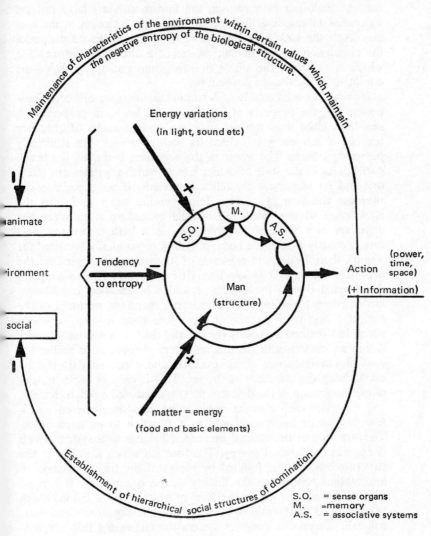

Figure 8

structure. And while the individual can and does succeed in being included in a human group, i.e. in realising his structural opening through inclusion in a sub-set, the human sub-sets have not yet succeeded in realising their opening through inclusion in the species. And the lack of a homogeneous species-structure precludes the circulation among human groups of a circulating information which is valid for the whole human group and not just for the dominant or dominated sub-groups.

It should also be clearly understood that what we call consciousness is not the property of one organ, the brain, in particular. A new-born child does not yet have any consciousness of existence because it has not yet realised its bodily outline: but it already possesses a brain. The whole of the organism, including the brain, participates in the state of consciousness which regulates our effective and no longer merely reflex behaviour. If we experimentally suppress the sensory paths afferent to the brain and thus the stimulation of our sense organs (the procedure is to place the organism in a state of weightlessness in a bath whose temperature is exactly that of the body, in a dark room totally insulated for sound), then despite the existence of its memorised experience the organism will lose consciousness. (It is necessary also to suppress the internal signals by inducing a stable biochemical equilibrium in the internal medium.) Consciousness, therefore, necessitates the perfectly organised circulation of information through a whole which has realised its bodily outline and thus its autonomous unity within an environment. In this set, every element must realise for itself the maintenance of its structure, and it can only do this by maintaining the structure of the set. This is only possible by way of the integration of the different *functions* secured by each element, and therefore each element needs to be seen simultaneously as a *structure* to be maintained and as a *function* to be encouraged. The structure is maintained because it belongs to a system which is open at the level of *energy*. The function which is linked to the structure can only be fulfilled by means of the free circulation of information concerning the finality of the organic set. If a meal goes on for several hours, the level of glucose in the blood, which has sustained the functional activity of all the organism's activities, will fall. There is a complex system for circulating this information. It must warn the liver, which releases glucose from its reserve glycogen; the brain is then alerted and told to look for a possible quarry in the environment to replace these reserves. The liver, muscles and brain have no expansionist aims in all this. They are

92

not trying to accumulate the greatest possible amount of reserves, to produce as much sugar as possible for the liver, as much mechanical power as possible for the muscles or the greatest possible flow of nervous impulses to the brain. They are not programmed to acquire the maximum power over others, but to act in such a way that the whole set of the elements which constitute the organism should be able to maintain their relations as a set, their structure in a given environment.

The circulating information is therefore linked to the finality which must be realised by the information-structure.

6

Information,
hierarchies of value and social classes

In my sketch of the historical evolution of domination in human societies in Chapter 3, I drew attention to the fact that human societies are in this respect strictly identical with animal ones. In every case the establishment of hierarchies is a result of the individual's pursuit of his biological equilibrium and his satisfaction. To achieve these aims he must dominate other individuals in the group. But in order to survive, the group itself must dominate other groups; and for this, it must appropriate the environment and exploit it as much as possible, so as to take from it the mass and energy which the group will use to increase its power.

In this latter sense, human societies are very soon distinguished from animal societies. In fact, as we have already seen, this exploitation of the environment depended on the quantity of information that the individuals were able to add to mass and to energy. A man who could shoot a bow and arrow was obviously less vulnerable than one who was still only able to cut flints to make a primitive pick. The former could dominate the latter at a distance, without engaging in hand-to-hand fighting.

As information became increasingly abstract, and with the advent of the industrial revolution, man was able to master energy and to treat matter in such a way as to manufacture considerable quantities of objects, thanks to the inventiveness of machines. To begin with this was simply in order to accumulate capital by selling these objects, capital being then as now the most effective means of domination for man over man and human groups over each other.

As long as these objects were produced essentially by the manual labour of the worker, the accumulation of capital was effected by the agency of surplus value, as Marx demonstrated, with the owner of capital retaining a part of the product of human labour which was not restored to the person who had supplied it. As machines became increasingly important in commodity production, manual labour became correspondingly less important in the production process. By investing, the capitalist used surplus value to appropriate means of mass production, machines. In this way he increased his power, because without machines the worker became ineffective; since these machines did not belong to the worker, it became possible to compel him to accept anything the boss wanted. For this reason the disappearance of private ownership of the means of production is a necessary (though insufficient) condition for the disappearance of domination.

But these machines, which are capable of supplying a considerable quantity of commodities in a minimum amount of time and of considerably increasing productivity, were "daughters of invention": [39] first of all in the sphere of the basic sciences, and then in their application to the industrial sphere and to the production of consumer goods. They were therefore also "daughters of abstraction". They were "programmed" by man, which means that by establishing their "structure" man turned them into depositories of information. Just as the worker's training was aimed at turning his nervous system into a depository of occupational information, which then found itself becoming a particular type of behaviour and of action on the environment (giving his movements effectiveness, precision and rapidity of execution in the production of commodities), so the machine became a depository of information leading to the production of objects on a massive scale. Its effectiveness was immediately much greater than that of man, for there was no limit to its power in terms of energy, and hardly any limit to its operating speed. It was less subject to errors, and although it needed maintenance, it did not put in wage claims. The worker might refuse to hand over to others (for their profit and domination) the information stored in his nervous system by his training, but the machine simply waited for energy to be fed in so that it could obey orders. And the worker was only there to "serve" the machine, not to give it orders.

So it was not the machines whom the capitalist had now to reckon with but those who supplied the necessary information for their invention, construction and utilisation: the technocrats. Among these — and this was a direct consequence of the increasing abundance of production — some were specialised in the administration of the production process, others in the circulation of products. In any case, when machines arrived, information became the important element in the production process, at the expense of the worker's labour power. This labour power itself had an energy aspect, a thermodynamic aspect intimately linked in the

95

human organism with the informational aspect acquired by training. With industrialisation, the energy aspect was to a considerable extent taken over by the machine. The informational aspect was also taken over, but the source of this information was the human nervous system. In short, the most important human contribution to the production process has become informational. It is man who is behind the information stored in the machines. And since it is machines that permit mass production and consequently massive profits, the supplier of information becomes increasingly necessary to the expansion of production.

What is then distributed in the form of economic and hierachical power is, of course, principally the informational content of the product of human labour. It follows logically from this that the "surplus value" which the "worker" surrenders at whatever hierarchical level he is situated is mainly *information*. The richer his labour is in specialised information and the higher the degree of abstraction attained in his training, the greater is the portion of information which he surrenders to the monster known as "the boss" and the greater also is the extent to which he is robbed. In this sense, the more "intellectualised" a job is, the more the worker is exploited, since in our times the amount of surplus value depends on the degree of abstraction in the specialised information which the individual is capable of handing over.

This is without doubt the reason why the big international firms have created their own technical schools or offer scholarships to students, from whom they then expropriate (under contract) the informational surplus value which they have paid for in advance. States do the same thing with their civil servants and their higher education. Things are already getting more difficult with the universities, although even in this case it is worth noting that the great seats of learning supply not only a lot of highly abstract stuff but a "culture" into the bargain, i.e. a way of seeing human life which fits in with the institutions of the dominant in such a way that those who will soon obtain economic and political power cannot call this power in question. The information supplied in these brain factories is therefore specialised in two directions, occupational and cultural, and the competition which takes place is based on conformity with the cultural framework of the dominant groups. This is the price of promotion in the hierarchy. Thus one can come across specimens who are often particularly effective at the technical level and completely stupid at the political level, since they are sufficiently satisfied with their domination not to search for a

96

clear understanding of its causes and meaning, and above all they don't want to jeopardise it in any way. If they have "made it", as the saying goes, it is simply because they are gifted; and the society that knows how to recognise their merits and give them power is therefore a just and equitable one.

The individual is only a recipient, an informational receptor who, in a secondary capacity, hands over to the purchaser something that is no longer labour power but a matrix of information which can be used to produce objects in greater numbers. The portion of information contained in what is now a "mechanofactured" rather than a manufactured object is constantly greater, and the human thermodynamic part less and less important; the "mechanical" work, the work that is poor in information, is increasingly done by machines. The latter is necessary for the invention and construction of the programmed machines which undertake the thermodynamic work in man's place. Finally, the automatised programme in man's nervous system gains in abstraction what it loses at the thermodynamic level.

It might be objected that we are simply talking about the individual handing back a gift of information that has been granted to him by society. But whatever the level of abstraction of the occupational information stored in a person's nervous system, it always comes from his socio-cultural niche in the surroundings. Whether his training is manual or conceptual, it is always simply a question of handing something back. It is only the "discoverer" who gives more than has been given to him, who supplies more information than was entrusted to him by his training. Indeed, one is forced to say that so far, under whatever régime, remuneration has taken the form of information not restructured by the imagination, i.e. automatisms, and that hierarchies are formed according to the degree of abstraction in which these automatisms are situated. The higher the degree of abstraction, the better the payment for the automatism. So far, creative imagination has never been hierarchically remunerated. It is easy to understand why. If this creative imagination is exercised in the domain of structures and not in that of consumer innovations, it constitutes a threat to the existing hierarchical and socio-economic structures and to the structures of domination. It is not, therefore, something that these hierarchies can contemplate; their finality is to conserve themselves as hierarchies. But to act in this way means to block evolution completely, and to encourage structural sclerosis by encouraging productivity in consumer goods, which is the basis of the structures as they currently are. We shall return to this question when we discuss the methodology of discovery.

I have stated that the most abundant source of surplus value is not manual labour, labour power, the thermodynamic activity of the worker, but the use of information in an increasingly abstract form — abstract, because this information is capable of shaping material, of "transforming" it (into machines) and "informing" it, and because these machines enable production (consequently profit, and therefore domination) to increase considerably. Why then, one may ask, are the technocrats not at the head of the revolutionary movement, against the power of the owners of capital?

There are several obvious reasons. The most immediate answer is that domination is no longer linked merely to the power of capital. Remember that the fundamental principle of a living organism is the pursuit of pleasure, which it obtains by domination. As long as this leads to two types of individual, the master and the slave, the oppressor and the oppressed, the dominant and the dominated, then the hierarchical distinction is simple, and antagonism easy. As soon as a complex hierarchical system appears, things are no longer the same. What gives a complex hierarchical system its solidarity is the fact that one finds those who are dominant and those who are dominated *at each level of the scale*. In such a system every individual is dominated by others but also dominates someone "smaller" than himself. Even the most underprivileged labourer in our social system will go home, bang his fist on the table and shout, "Bring me my soup, woman!" and if one of his children misbehaves he will hit him. He will have the feeling that he is master in his own home, that he is obeyed, respected and admired; and the child will make his father the ideal for his own ego, even at that tender age. This domination within the family will often be enough to fulfil his desire for self-satisfaction. On the other hand, when he leaves his house he will find people who dominate him, those who are positioned immediately above him in the hierarchy of the degree of abstraction of occupational information. And like the chimpanzee in a position of submission to the dominant chimpanzee, his whole nervous system will be in a topsy-turvy state, secreting in a disordered way, for in our modern societies flight is impossible. He has to submit. He can no longer do battle, because he risks seeing his means of subsistence taken away from him. The result is that he suffers biologically every day; he has a sickness, a sense of malaise. Submission has nothing but drawbacks. Work which is "parcellised" and makes each individual closely dependent upon others is no longer felt only as alienation. Whereas palaeolithic man was a real polytechnician in the context of the technology of his time, modern man is incapable, whatever his technological level, of supplying his own basic needs by himself. What modern man feels as alienation is his inability to decide his own destiny, his inability to act upon his environment with an act of gratification. But from another point of view, this lack of decision-making power gives him a feeling of security. He knows that he has little chance of dying of hunger and that certain responsibilities are spared him. His large informational deficit is a source of distress, but meanwhile he puts his confidence in those who claim to know and act for him. In our developed societies, therefore, man's behaviour contains a curious mixture of unsatisfied desires on the one hand (which stem from his drastically reduced possibilities for taking acts of gratification on the environment), and of security on the other (because of his participation in a social whole which makes decisions for him and satisfies his basic needs). We shall need to discuss the question of power and democracy at length. But let me remark for the moment that this mixture of the satisfied and the unsatisfied is, for me, the source of our social "malaise".

The farther removed the level at which decisions are made the more abstract it becomes, the more likely is it to be obscured from view and overlooked. But in reality man's gratification, like the alienation he suffers from, is to be found in his immediate surroundings, in that part of his environmental niche which he can touch every day with his hand, and whose structure and causality he can find out quite simply. He tends to shift the responsibility for the gratification or suffering that he gets from his surroundings on to those levels of organisation which he has only an abstract idea of; he returns in modern times, as it were, to the first man's mythical tendencies concerning the gods. The modern gods are called Liberty, Equality, Democracy, the State, Social Classes, Power, Justice, Political Parties, etc., and their priests — whether clumsy or skilful, despotic or benevolent — are called Governor, Chairman of the Board of Directors, Bourgeois, Technocrat, Bureaucrat, Boss, Manager, Official, etc.

Modern man is deceived by some of these gods and their intermediaries, and sometimes tries to improve his fate by changing his religion. But he never questions the hierarchical system or its causes (we would say its behavioural factors), and he still has not understood that merely changing the pieces on the chess board does not mean the disappearance of the squares, i.e. the underlying structure of behaviour.

So we can begin to understand why the technocrat is very rarely a revolutionary and why he does not seek to do away with the established power or, more precisely, the hierarchical structure. It is this hierarchical structure that enables him to obtain gratification.

The activity of the social set is basically directed towards commodity production, which is a function of the invention, creation and use of machines; and since these in their turn are a function of the degree of abstraction of occupational information, it is easy to understand why a choice place in the hierarchy is reserved for the technician. He ascends in the hierarchy to the extent that he proves capable of handling information in an abstract manner rather than a thermodynamic one, for the higher the degree of abstraction, the more generalised will be the use of that technician and the more his productive effectiveness will be appreciated. But it is essential to understand that the "power" which he thus acquires is strictly limited to the process of production. There is no reason whatever, in principle, why it should be linked to any political power, because it is not based upon any political knowledge. "Political" is a word that we still have to define.

I have already emphasised that the handling of information is a particular capability of the human species, and that palaeolithic man immediately put himself in a different category from the animal species which had preceded him because he "informed" or "gave form to" inanimate matter by making the first tools. This particular quality in itself is not sufficient to give birth to hierarchies. That is why I have continually pointed out that the ladder of hierarchies ascends according to the degree of abstraction in the occupational information handled: information passes through all the levels, starting with the most concrete, that of the labourer, through the more complex level of the artisan, and becoming increasingly abstract with the engineer, the technocrat and the bureaucrat in general. The result is an infinite number of hierarchical levels which proceed gradually from the manual worker to the intellectual.

Where does the proletarian end and the bourgeois begin in this hierarchical scale? Marx defined the bourgeoisie on the basis of its private ownership of the means of production. Capitalists today will tell you that capital and the means of production are less and less the property of a few particular people and now belong to a great number. In the socialist countries they have even become the pro-

100

perty of the state, that is to say — in principle — the property of the collectivity; but in spite of this, hierarchical systems and the alienation they give rise to can hardly be said to have disappeared.

As long as hierarchical systems of value exist, the full "flowering of the individual", as it is called in election manifestoes, will only be a myth. In a hierarchical value system, as we have seen, every individual is dominant over some people and the victim of domination by others. It is therefore impossible for him to "flower". The transformation which will undoubtedly be the last to take place is the abandoning, by each individual and at each level of this hierarchical organisation, of the psycho-familial type of paternalism towards the members of any class he regards as inferior, and of childishness towards those whom he considers as his superiors (bosses or institutions) and who afford him security while at the same time preventing complete self-gratification. In my opinion, the particular strength of some hierarchical systems which are highly structured, such as the army, the legal system or hospitals for example, is due not so much to their hierarchical structure itself, which one tends to think of as terribly constraining, but rather to the fact that every element in the system, whatever his level in the hierarchy, is indoctrinated with the idea that he is part of an élite which is different and superior to all other élites because its "ideals" are higher. Their strength is due to the fact that desperately feeble value judgements are raised to the rank of an ethical code, and to the fact that the individual is gratified by this in itself. The uniform, *esprit de corps*, all the business of buttons, caps and berets turns the individual into a member of some master-race, and makes him accept his total alienation from the hierarchy without even asking himself what this hierarchical set-up is that he belongs to. It is this determinism which appeals to the most primitive functions of domination in the brain of a reptile, to the narcissism with which born, to the profligate colour display in the plumage of male birds during their courtship displays, to our subcultural automatisms of the crudest sort: this unconscious determinism is what we call "freely accepted discipline". These strongly hierarchical organisations appear to be generously disinterested. In a world dominated by profit and commodities, their members march in step, heads held high, without lowering their eyes to the money they might be getting elsewhere, poor but honest. But ask the lieutenant whether his ideal is not to end his career as a captain, whether promotion in the hierarchy is not the motivating factor that dominates his behaviour. You will hear them say — and you don't know whether to laugh

101

or cry — that it is their vocation to "command". The vocation of discovery, or at least of exercising their imagination, never seems to come into their heads and, because hierarchies usually have little time for improvisation, their ideal is limited to carrying out drill. And their advance up the hierarchy depends upon their submission to this drill.

Drilling has only one aim: to protect the socio-economic system (in reality, the hierarchical system) which pays them. There is a lot of debate about the "professional army" versus the "army of the nation". But it really makes little difference, for it hinges on a false problem. In both cases it is simply a question of protecting a social structure. And all social structures so far have been hierarchical value structures.

It is only when the hierarchical levels no longer offer a sufficient number of intermediary steps, and when the functional classes in the social body are few and strictly segregated, that the risks of violent explosion are likely to arise. Where gratification through social advancement within the production process is difficult or impossible even though people submit to the rules of institution-alised domination (exams, competition, etc.), aggressive reactions are probable. They are quickly controlled, often by the use of armed force, which as a rule puts itself on the side of the dominant. who obviously defend the existing social structures. Such hierarchical structures forbid any circulation of information and thus also any human cohesion as a group, and they perpetuate domination. Sooner or later the result is a "crisis" or explosion, which can be interestingly contrasted with the malaise that results from establishing hierarchical levels with a clever mixture of satisfied and unsatisfied desires. The crisis appears in the form of violent antagonism between closed structures, as in war.

We can thus see how the change from distress to malaise or from distress to crisis takes place. We should not forget that satisfaction is obtained basically by an act of gratification on the environment. If that is *impossible,* a crisis is *possible*; in the case of the individual we call this "aggression" and in the case of a whole population we call it either "revolution" (when the conflict takes place within one social organism, between two national sub-sets of people) or "war" (when it takes place between two national structures). But it may be that the action is neither fully gratifying nor fully realisable, as is the case in societies which have many levels of hierarchy and whose finality (which is expansion) does not coincide totally with that of the individual. If the mass media have not sufficiently motivated the individual towards finding his gratification in a goal which, even though he benefits from it, lies outside himself (in production, for example), then the malaise emerges. Whatever the case — distress, social malaise, crisis or war — the source must be sought in the "closure" of the information-structure at a certain level of organisation.

For example, the biological disequilibrium which causes the sensation of hunger may, at a certain level of intensity, be experienced as a malaise; if it is possible to get food, it will disappear. If another closed set — an individual or a group of individuals enforcing the law — forbids or delays the appeasement of hunger by creating automatisms, a malaise will result. This malaise will, however, itself be able to serve as gratification if the automatisms are strong enough to turn it into a long-term gratifying finality for the person experiencing it. This will then be a sacrifice joyfully accepted for the sake of the future, in this world or the world to come. But man's drives and automatisms are further complicated by the imagination. Malaise can be caused by the imagination: by a happy future imagined but not fulfilled, or by a painful future imagined as possible, even if it never happens. At each level we once more come across the principal mechanisms that we have already noted with regard to distress. And as in the case of distress (see page 49), the informational deficit plays an important role.

Should we speak about class structures?

It seems to me difficult to understand the phrase "power to the workers" which is so often used. In present-day society, anyone who can live without working is very lucky. Very lucky and very rare. Work can be well paid or not so well paid, in any case, badly paid work is not the only kind that exists. Does work become pleasure the moment it is well paid? And what does one say about a person whose pleasure is work, whatever the payment? Such cases are rare but they exist. It is something I know about. Would it not be more precise to demand: "power for those who haven't got it"? But does this mean consequently that those who possess power should lose it? The problem is badly posed. To give power to those who haven't got it does not make it necessary to remove power from those who have it. The desirable objective is to generalise power, for then there will no longer be any power. The mistake springs from an excessively narrow conception of "social classes" and of the classic opposition between capital and labour.

In any organisation whatsoever, the analogy by which individuals are grouped is one of function, in actual fact. One generally associates them by an analogy of hierarchy — management, staff, workers — a hierarchy in which power is a regressive factor in decision-making and in the smooth running of the enterprise. However, alongside the hierarchy of value which satisfies the power instinct, there exists this fundamental hierarchy of function, which I have preferred to call functional "levels of orgainsation", to remove any value judgement.

Indeed, although I have spoken so far about hierarchies of both value and function, this is simply an aid to understanding, for in reality every hierarchy is a hierarchy of value. By contrast, the organisation of an individual body or a social body displays "levels of organisation". Each higher level embraces the level of complexity that precedes it; it does not give orders to this preceding level but informs it, by means of the "circulating information" which I referred to earlier and which I distinguished from the "information-structure". What are the "classes" of elements in this kind of individual organism? We know that there are different functions, all informed by the finality of the set on which the metabolic activity running their "occupational" work depends. In other words, there are *functional* classes, of many kinds, each co-operating with the activity of a large system (the nervous system, the endocrine, cardiovascular, respiratory, locomotive, digestive systems, etc.), and each co-operating in the activity of the organic set within the environment. These functional classes have nothing to do with the hierarchical classes which appear in "class struggle". But when we discussed the idea of information in human sociology, we saw that the hierarchical rungs are staggered to such an extent that it is impossible to know at what point one leaves the proletariat and enters the capitalist class, and impossible to know by what criteria an individual can be assigned to one class or another (unless by virtue of a state of mind or membership of a party). It is therefore questionable whether the notion of class as it was experienced and understood at the beginning of the century still has any more than an affective reality.

On the other hand, the notion of hierarchy as we have defined it certainly corresponds to a reality. It corresponds to a functional characteristic in the mammalian brain in general: the pursuit of domination. And in the human species, it corresponds to the degree of abstraction which "information" introduces into human labour.

Therefore when I speak of social classes I shall be referring to *functional* classes: that is to say, to all the individuals in a social organism who perform the same function or similar ones. It is only the consciousness of this kind of class, and thus of its indispensability (and the indispensability of the *other* functional classes), that will enable us to achieve the "dignity of the human person" which electoral addresses always speak so much about. One has no more "dignity" when one is rich than when one is poor, with a high position in the hierarchy or with a low one. On the contrary, when one is rich and holds an exalted place in the hierarchy, one believes that one has more "power". One attains pleasure more easily by domination. Which is the same thing as saying that the dominant will always seek to keep their power and leave to the dominated "their dignity", which they couldn't care less about. What the dominated aspire to is not a dignity whose possible use to them they are (correctly) unable to see, but actual power. As long as power is linked to specialised information, which makes promotion in the hierarchies of income and value possible, human societies will not be able to surmount the thermodynamic stage of productivity, since directly or indirectly these hierarchies are simply based on the quantification of the individual's specialised informational participation in commodity production.

It seems to me that a lot of people have let themselves be conditioned by certain stereotyped automatisms of thought which are linked with the expressions "class struggle" and "classless society", and with the definition of the capitalist class as the class which privately owns the means of production and of the proletariat as being characterised solely by its "labour power". Surely one cannot fail to see that it is not sufficient to suppress private ownership of the means of production in order to arrive at a classless society and to end class struggle. The hierarchies of value are not only linked to the possession of capital and of the means of production. Power today is a function of specialised information and it is this above all that enables people to establish themselves in positions of domination. As long as hierarchies of value based upon specialised information are not suppressed, there will be those who dominate and those who are dominated. On the other hand, if a hierarchy of function comes to be established, there will be as many social classes as there are established functions, and one and the same individual will be able to belong to several social classes at the same time, in several different institutions, pursuing several different activities (consumers' associations appear to be giving birth to a new class of this type).

As long as hierarchies of value persist and as long as they are based upon property, through possession of the specialised information acquired by a manual or intellectual training, the dominated will try to wrest for themselves a false power, namely the power of consumption. There is no end to consumption, and it will never be possible to establish a real equality of opportunity and power on the basis of consumption. The real power which the dominated require is not so much that of consumption as that of participation in decision-making. And for this, the dominated must acquire generalised information, not just specialised information.

The success of marxism proves beyond doubt that it supplies a general grid with which to decode the relations of production. It is already generalised information in a sense, but unfortunately its foundations were laid in the last century, in ignorance of the ideas of information and behavioural biology. So it is partially generalised information, a stage in the growth of knowledge, just as the ideas that I am trying to disseminate are another stage — a more "general" one, certainly, since the ideas are enriched by knowledge that has been acquired since the last century, but necessarily still a very incomplete one. The danger of a grid, however effective it may temporarily be, is that it is liable to suffer from a hardening of its conceptual arteries; this was the case with Aristotle's grid, which before the advent of Marx and Freud congealed millions of minds with an incomplete conception of things. The person who finally possesses (or thinks he possesses) the key to the most beautiful human dreams because he feels secure and has succeeded in covering up his anxiety, is thus prevented from setting out once again on the long march towards the unknown, towards distress, uncertainty, chaos, the dangers and necessary stages that precede the creation of new structures.

Since people in industrial societies have never been told of the existence of any kind of education [*formation*] other than occupational education, nor of any information other than occupational information, they accept the occupational hierarchies as inevitable, for the society's finality is solely occupational, being centred on commodity production. It comes, then, as something of a surprise that the people enclosed within these hierarchies and piled high with consumer goods are, despite all this, the victims of a certain malaise.

7

Consciousness, knowledge, imagination

Elsewhere[40] I have contrasted these three words with three others which are as dangerous as they are unrealisable: "liberty, equality, fraternity".

"Liberty"—when one has lost it, it is then one most appreciates it, or so it is said. People think about the loss of free speech, loss of free movement across frontiers, loss of religious liberty, of the right to join political parties, etc., in the socialist countries today, and forget what is happening in Chile, Brazil and elsewhere. But take the pedigree cat, who is cossetted on soft cushions and caressed by his master, totally automatised in his behaviour as regards food, excretion and play: is he freer than the alley cat? Or, to turn the argument the other way round, is the alley cat, which at every moment of the day has to think of his survival and look for food and shelter and is at the mercy of every change in his environment, more free than the pedigree cat? Was palaeolithic man more free or more alienated than modern man? I don't intend to develop this debate about freedom here. I want simply once again to emphasise that alongside the more or less alienating determinism of the environment there are also the various forms of determinism (rarely regarded as such) in the mechanisms of our central nervous system, and that these are much more alienating. A prisoner within his prison walls is free to dream; and a managing director, apparently free to go where he wants, will do so only in obedience to an alienating myth of property, profitability and production. The pursuit of domination involves us in the most primitive mechanisms of our central nervous system, with the inevitability of a policeman's handcuffs.

What we call freedom is the ability to carry out acts of gratification, to attain what we set out to attain without colliding with someone else's objective. But the act of gratification is not free. In fact it is entirely determined. The absence of freedom, then, is a consequence of the antagonism between two behavioural determinisms and the domination of one over the other. Looked at from this point of view, freedom consists in the creation of cultural

107

automatisms of such a kind that the behavioural determinism of each individual would have the same finality, but would be situated outside him. One can see that in a hierarchical system of domination this is impossible except in periods of crisis, whatever the socio-economic régime may be.

What we call freedom amounts, generally speaking, to the ability to respond to our primitive drives, which are already considerably alienated by the socio-cultural automatisms, prejudices and value judgements of the social group and of the epoch in which we are integrated. Liberal societies have succeeded in convincing the individual that freedom is to be found in obedience and submission to the rules of the hierarchies of the moment, and in the institutionalisation of rules that must be obeyed in order for him to rise in these hierarchies. And not much is heard about the lack of freedom for the worker or for the people of the underdeveloped countries, nor indeed about the lack of freedom for all human beings chained to the laws of production. It is a poor kind of freedom which is content to remain ignorant of the determinisms governing our social behaviour. Freedom begins where knowledge ends. Before this point freedom doesn't exist, since our knowledge of laws compels us to obey them. After this point, it only exists by virtue of our ignorance of the laws to come and the belief that because we are ignorant of them we are not subject to their dictates. In reality, what we might call "freedom", if we really are set on keeping this expression, is the very relative independence that a human being can acquire by discovering, partially and gradually, the laws of universal determinism. Then, and only then, he can envisage a means of using these laws better for his survival; this causes him to penetrate another form of determinism, at another level of organisation of which he was at that point ignorant. The role of science is to penetrate ceaselessly into a new level of organisation of universal laws. As long as man was ignorant of the laws of gravitation, he believed that he was free to fly. But Icarus crashed to the ground. When the laws of gravitation became known, they enabled man to go to the moon. In doing so, he was not freed from the laws of gravitation but was able to use them to his advantage. How can we be free, when an implacable explanatory grid forbids us to conceive of a world in any way different from the one imposed by the socio-cultural automatisms which it prescribes? How can we be free, when what we claim to be a choice between this or that is actually a result of our instinctive impulses, our pursuit of pleasure through domination and our socio-cultural

108

automatisms, determined by the environmental niche in which we happen to be? And how can we be free, when what we possess in our nervous system is simply our relations with others, internalised? An element is never separate from a whole; an individual separated from any social environment becomes a "wild child" and will never be a human being. An individual does not exist outside his environmental niche, which is like no other and which entirely conditions him to be what he is. How can we be free when we know that this individual, an element in a set, is equally dependent upon more complex sets which envelop the one to which he belongs? When the organisation of human societies up as far as the largest set, which constitutes the entire species, develops by levels of organisation each of which represents the control of the servomechanism regulating the level below it? Freedom, or at least creative imagination, is only to be found at the level of the finality of the largest set, and even at this level it doubtless obeys a cosmic determinism which is hidden from us, for we do not know its laws.

Thus "freedom" is not the opposite of "determinism". The latter can no longer be conceived as it was at the end of the nineteenth century as a determinism of linear causality, with a cause producing an effect. Socio-political "analyses" often still refer to this infantile kind of determinism. Effectors whose structure we often do not know about give rise to multiple effects following the action of factors that are also multiple; they are far from all being identified and measured, and they are themselves controlled by the feedback coming from the effects. They constitute very complex systems; because we lack knowledge of them, we are unable to speak about freedom or about their aleatory nature, but only about our own ignorance.

"Equality" is an empty concept which has nonetheless motivated people for centuries. Equality conceived as "identity" is contrary to common sense. But as soon as one admits "difference", and one cannot avoid doing so, how can one then keep the concept of equality? One can see, however, the drives which have motivated it: the search for pleasure and for biological equilibrium on the part of each individual in the social situation, that is to say in a situation which so far has always been hierarchical. How is it that ideologies can still mobilise masses of people with the concept of equality while they still desperately cling to hierarchies of power, income, knowledge, etc.? There is the issue of "equal opportunity".

But opportunity for what? For education? Education allows people to climb the hierarchical ladder; technical and occupational education, which is more or less abstract, allows one to dominate in the world of commodities. Education also allows access to consumption and respectability. Why not, in such a system? But then the family system, the bourgeois Oedipus, the environment niche, must also be involved. And even supposing one could make all these environmental niches uniform, man would still have a "talent" for ascending in the hierarchies of technique, consumption and fame. Where would this talent then come from, if not from his experience in the environmental niche? It would come from what is innate, from a fortuitous encounter between an egg and a spermatozoon — from what we may call "chance" (the genetic combinatorial is subject to so many factors that one soon gets lost among them). This is an inescapable dilemma: if one makes the sociological opportunities of access to technical and occupational information uniform, then either one falls into the basic injustice involved in "inherited talent", or one gets individuals who are all identical in their behaviour, motivations and socio-cultural automatisms, and even in their imaginations. So equal opportunity, much as one may desire it, amounts simply to the capacity for being content to be in one's own shoes rather than someone else's. But this is only possible outside of hierarchical systems, since it is hierarchical systems that institute economic inequalities and inequalities of domination and gratification.

But if equality cannot exist in the living world, this does not mean that power ought to be distributed hierarchically. Equality only exists in the indispensability of the functional classes, for indispensability is an absolute criterion — it is the only effective basis for equality that works. But then it is a question of the equality of *political* power — and for a functional class, not for an individual in relation to other individuals.

In a living organism, no cells, no organs are free or equal. One works more or less than another, and needs to consume more or less than another. When they are free or equal, this means that there is cellular anarchy (cancer), or a malfunctioning of systems which are incompatible with the survival of the whole. Moreover, their only interest in individual freedom lies in fulfilling their "desire" (i.e. the maintenance of their structure) by means of the coherence of all their partial finalities with those of the whole set. The finality of the set cannot help being theirs too. No individual, no cell is indispensable to the smooth running of the set. But on the other hand, the union of many individuals performing the same function in organs, and the union of certain organs in systems, is *indispensable* to the functioning of the organic set. So when the passage from one level of organisation to another occurs, when there is inclusion of one set in a larger set, the "freedom and equality" of the elements in this set loses its meaning, whereas the indispensability of the sub-sets acquires a meaning. The closed information-structure of the individual is unconscious of the relations which unite it with its environmental niche, and the closed information-structure of human groups and human societies is also unconscious of them. We still carry empty concepts about with us. Liberty will only commence when every individual is totally given over to the finality of the species, with the latter no longer finding antagonistic organisations in the biosphere which are capable of making him desire otherwise.

As for "fraternity", it is impossible even to undertake a rational critique of it; merely by looking at the facts, we can see to what an extent this abstraction often abuses people. At all events, there is no point in preaching it ideologically as the pursuit of a paternalistic humanism, for the truth of the matter is that men have not been brothers since neolithic times, except within the framework of the Oedipus triangle which has spread to all societies: daddy, mummy and me. Fraternity cannot only be a fine sentiment. As long as it remains that, it will be totally ineffective, since the unconscious determinisms that govern the domination and exploitation of man by man (which always find convincing alibis in rational discourse) are very strong. Fraternity can only be envisaged with knowledge of the mechanisms governing the establishment of the "human personality" and the unconscious determinism of behaviour. It demands the suppression of hierarchies: without this, it will never be more than a covert paternalism. Instead of having the words "liberty, equality, fraternity" inscribed on public buildings by a bourgeoisie for whom they signify the domination of shopkeepers or industrialists, societies of the future will write the words "consciousness, knowledge, imagination": consciousness of determinisms, knowledge of their mechanisms, and the imagination that enables them to be utilised for the survival of the entire set of human beings living on the planet. Consciousness and knowledge are the foundations of tolerance, and this itself is close to fraternity.

The imaginative effort will perhaps one day justify the term

111

"complementarity" which is infinitely richer in evolutionary possibilities. A "complementary" set has no need of equality or freedom, since it is *indispensable* to the realisation of a greater set in which antagonism and domination disappear.

Men and women, instead of each claiming for his or her sex an "equality" which nature itself has not been able to induce (it is quite simply incomprehensible at the level of the spermatozoon or the egg) could perhaps seek a complementarity of this kind, founded upon the indispensability of the two sets which they represent. It is at the level of the capitalist family, which has been adopted universally by the most resolutely socialist régimes, that the most fundamental reform is perhaps most necessary. When will self-management be possible in the family group (if, indeed, there still has to be a family), with no hierarchy, no domination, and full knowledge of our primitive drives and socio-cultural automatisms? The answer is: when there is full consciousness of the possible openings which alone are capable of transforming an inexorably closed system into a system that opens wide on to the world.

This critique of the concepts of liberty, equality and fraternity is not a mere stylistic exercise. What we actually have is a set of words with a very imprecise semantic content into which each person brings his own experience, which is never that of his neighbour. Consequently, few words are so capable of rousing elementary emotions in crowds and guaranteeing the permanence of the existing hierarchical systems.

"Liberty" is linked to the *decisions* which are indispensable for the maintenance of domination. One assumes that when a decision is taken there is full knowledge of the reasons for it, and that it is therefore linked to *knowledge*. We are, of course, talking solely about technical knowledge, which has made a fortune for the technicians: it is a knowledge which is generally quite separate from the conceptual sets which comprise this technique. The acquisition of technical knowledge enables one to acquire a favourable *hierarchical position*. So decisions never go against the maintenance of the hierarchical system which gratifies the decision-maker. This means that decisions are not *made*; in a given hierarchical socio-economic system, decisions are quite simply *obedient to the system*. They are therefore not free decisions: they are determined.

"Liberty" is also linked to "responsibility", the heavy burden which at first sight seems to fall on the shoulders of managers and bosses. Responsibility is the counterpart of the domination enjoyed by those on whom it falls. But if there is no freedom to make decisions, there can be no responsibility. At most one might say that the performance of a certain function demands a certain level of abstraction in technical knowledge and a certain amount of occupational information. If one possesses this experience, then the decision is forced upon one, it is spelt out by necessity. Or if there are several possible choices to be made, the solution ultimately adopted belongs to the domain of unconscious motivation or socio-cultural experience.

It is quite exceptional for a solution supplied by the imagination, which is moreover itself motivated by the above factors, to contribute anything new to the tried and tested automatisms. It is the "innovator's" job to express it, but he comes up against the hierarchical systems, which cannot admit novelty if it is not saleable. He therefore stands little chance of rising in the hierarchy. Hence the rarity of motivations leading to innovation, for this is rarely gratifying in an established hierarchical system.

It might seem that recognition of the non-existence of liberty, decision-making and responsibility would lead to the disappearance of any gratifying motivation. There would no longer be any reason for trying to rise in the hierarchy. There would be no reward for values that do not exist. There would no longer be any hierarchy either. The world of production for production's sake, that is to say production in order to obtain domination, would crumble. Might not human motivations other than those of rising in the hierarchy then emerge?

A whole conception of human behaviour, which up to now has provided the happiness of those who dominate, vanishes once these worn-out concepts are demystified. The same arguments apply to the notion of "will", which fortifies people. Does this represent anything more than the power of motivation, and if so, is it not essentially a consequence of the hypothalamic function? The more the satisfaction of a need is felt to be indispensable to survival and to biological equilibrium or "happiness", the stronger will be the motivation (that is to say, the will) to satisfy this need.

Socio-cultural training, from generation to generation, demonstrates to human beings from infancy that effort, labour and will are the foundations of social success, the key to promotion in the hierarchy. What part does this training play? The ideal of the ego cannot establish itself in this context unless priority is given to "the will". And finally, when this will comes up against the barrier

113

of social inhibitions, when the will to achieve the act of gratification collapses, then it is rare for the *imagination*, which turns away from the conflict to find an unforeseen solution, to be valued as highly as the will once was. Social groups have no use for the imagination, which upsets things, whereas the will, in a determined hierarchical framework, reinforces the hierarchical structure of the group. But the "will" is at the mercy of a small passing deficiency in the subrenal corticoids, and of a depression in the synthesis of the cerebral catecholamines; and this depression, which results from an insufficient organic reaction to the inhibition against the act of gratification, sees the whole edifice of will disappear for (mere?) biological reasons.

Much the same might be said of courage and cost of the qualities so deeply prized in warlike nations, for they automatise individuals for the benefit of their particular structure of domination. I hesitate to say that there is no such thing as courage in itself, or will in itself. Surely the important thing is to discover a motivation whose finality is the same for the individual as for the species: a motivation which is sufficiently open to both these poles, and sufficiently integrated at the level of consciousness for life to be intolerable if the motivation is not satisfied. You will then have will and courage into the bargain, particularly if your subrenal glands are in good condition.

Finally, why pass a value judgement on those people who make an ideal image of themselves and who try to realise it and share it with others? Is this not one way of giving oneself pleasure? Do I not prefer, in the morning when I am shaving, to see a man's head in the mirror rather than that of a chimpanzee? The fact is, of course, that our understanding of human beings is poor, even poorer than our understanding of chimpanzees.

8

Democracy and the idea of power

Democracy is a dangerous word; it is used without a precise indication as to which of its possible meanings is being used. Government by the people. What is the people? Is it the total set of human beings living in the same country? Is it the majority? The total set of individuals governed by an oligarchy? The least well-off sector? The least educated sector in a nation, and if so the least educated in what? Let us agree that it is a little of all these at one and the same time, but that it is pre-eminently government by the greatest number freeing themselves from the oligarchies. It would be a remarkable state of affairs if the oligarchies governed for the people and not for their own well-being first and foremost. But even supposing they were to do just this, it is precisely this kind of government, even when it is for their own good, that the people today reject. They don't want to delegate their power to anyone to act on their behalf, and indeed one can only congratulate them on this. For history teaches us that even in those countries where private ownership of the means of production has been suppressed, ignorance of the bio-physiological bases of behaviour ensures that every benefactor of the people who does more than speak in the name of the people and gets involved in action will very quickly find that he is acting for himself, or for concepts which only exist in his own head and which are not for the most part those of the commonalty, of the people. Again, there is fusion between the legislative (informative, one might say) and the executive branches. The people very quickly find themselves exploited by those who possess "information" and who use it to secure their "power".

But information is necessary for effective action; the people cannot act, since they are either simply not informed or, more seriously still, informed in a one-dimensional way, with a slant which ensures the maintenance of the hierarchical structures and the structures of domination. This is the situation in capitalist and existing socialist régimes alike. As long as information is in the hands of a particular set of people, as long as it is disseminated

115

from the top down after being filtered, and as long as it percolates down to the recipient through a grid imposed by those who, in order to meet the needs of their own domination, do not want this grid to be challenged or transformed, then democracy is an empty word: it is the counterfeit coinage of socialism.

In order for information to be able to spring up everywhere and not always to flow from the same tap, total freedom of expression (I don't say of action) and total freedom of dissemination for what has been expressed is clearly indispensable. It is a tough job as things are for a new idea to get heard, quite apart from the fact that freedom of expression is muzzled simply because the new idea is not written into the preceding nerve patterns. The new idea causes distress because of the informational deficit. It is not enough merely to send out a piece of information, it is also necessary to have a system sensitive enough to receive it; usually, this piece of information is like a new television channel which no set can actually pick up. It goes out on the air without being transformed, without being translated into a picture. But essentially, even if it can be picked up and translated into a clear language, it is also necessary to have viewers to record it and fix it in their memory: people for whom it constitutes an additional experience, an enrichment of the material on the basis of which they operate their own individual associative systems. So people must have free time to receive and process it. This free time will be subtracted from the time devoted to productive work, and consequently production will suffer.

This is a fundamental idea. The existence of socialism depends on the amount of time granted to the people (to the greatest number of people). Whatever the level of technical education of the individuals who make up "the people", production and growth must if necessary suffer (and probably will) in order for them to become informed.

I have already pointed out that in my opinion there are not a limited number of classes — the capitalist class, the proletariat, the tertiary sector — founded on ownership or lack of ownership of the means of production, but an infinity of social classes which I have called "functional" classes. The false distinction between a limited number of classes is a result of establishing economic, sociological and political concepts without drawing the distinction between information and thermodynamics.

It is very difficult to make "manual" and "intellectual" workers

116

cohabit in this kind of simplistic framework. The result is that hierarchies and therefore inequalities in power arise, founded on something that is overlooked or simply not take into account: namely, the quantity of specialised information handled by an individual. Alongside this overloading of the informational and technical hierarchies, an attempt is made to establish equality at the thermodynamic level of consumption; but this remains only an aim, since the informational hierarchies are preserved by the very fact that they are overlooked, and they remain what they already are — hierarchies of income and occupational power.

Clearly, in all known systems what the individual gets paid for in income and power is simply the *information* which is introduced into his nervous system and which he hands back to society in various thermodynamic forms.

From all this we can understand that the idea of democracy follows on from the definition we have given to "the people". Since modern societies consume more and more specialised information, and less and less mechanical human labour power, the law of supply and demand results in the establishment of economic hierarchies and hierarchies of occupational power which are based more on specialised information than on mechanical human labour (which carries only a small informational content). The length of time spent on schooling increases; and there is occupational retraining, that is to say, occupational life is recharged with information. If "the people" represent the occupationally less informed mass of the nation, it is guaranteed that in such a system the people will not be able to conquer political power. Taking the argument to the point of caricature, one could even imagine future societies in which the human masses, occupationally uninformed and consequently useless, would be paid for doing nothing because work would be almost completely automated. They would be granted an average *economic* power as a compensation for their total surrender of *political* power to individuals who are occupationally better informed and hence more useful in the creation, programming and control of machines and in the production of commodities.

But one could imagine, by contrast, future societies in which "the people" would be very broadly informed in this occupational sense, and in which the main effort would be directed towards training; this would begin increasingly early in life and would become increasingly conceptual and abstract. As long as hierarchies are still being established, according to the degree of abstraction

117

in occupational information, and as long as the finality of the social set remains commodity production, we may be sure that democracy will be a mythical hope, a word, and not a practical reality.

As information in general "is only information, and is neither mass nor energy" (Wiener), occupational information is only occupational information. So on the one hand it has no reason to reserve all political power to itself; and on the other hand, politics can hope one day to serve some other purpose than the control of production. Occupational information cites its "desire to safeguard human happiness" as an alibi. Human beings are made in such a way that they are unable to find general happiness in a system of hierarchies, because all hierarchies are simply the expression of domination: and this human make-up is the final fruit of a long and complex evolution in which all the integrations of centuries are summarised. But if the finality of the human species remains labour on the production of consumer objects, we may be certain that after the domination of hierarchies founded upon the ownership of capital (or replaced in some cases by the domination of bureaucratic hierarchies which organise production and serve as guardians of the social structures), a technocratic form of domination will emerge, founded on the degree of abstraction in occupational knowledge. Hierarchy will replace hierarchy, and the only change will come from the gradually increasing component of specialised information involved in establishing this hierarchy.

It seems to follow from what we have just said that it is necessary to separate political power from occupational power. It is true that occupational power is deployed within an institution of limited size, as the power of domination which specialised information has over thermodynamics But even at a high level of organisation, at the level of the integration of firms into industries and of industries into the nation as an organic set, the classes of "political power" spring from the establishment of classes of occupational power at the lower levels of organisation. The occupational hierarchies associated with specialised information thus extend to hierarchies of political power, and one must therefore recognise that the occupational power associated with specialised information is fused with political power of the "big men", which is no longer based merely on the ownership of capital.

If we take into account the differing proportions of more or less abstract occupational information in the nervous systems of the individual, and the "functional" hierarchies that result therefrom within any enterprise, the question then poses itself: is the maintenance of the "power" of hierarchical occupational domination inevitable? Because any single individual is ignorant of the sum of information acquired by another (a sum which perhaps finds no day-to-day expression in what he does in his occupation), he tends to minimise the range of this power. The same thing happens among the scientific disciplines: each individual in any one discipline tends to underestimate the knowledge of a person belonging to another, because he does not know the other discipline and his own personal "grid" seems to him adequately to account for the whole of the surrounding terrain. Consequently ignorance always contains a value judgement about the effort needed to acquire a sum of information and to debate its operational effectiveness. This debate is all the more generalised among individuals in so far as the system, as we have seen, links power with occupational knowledge. Someone wants to challenge a power by which they are oppressed because it opposes their individual desire to dominate and the pleasure which they might enjoy by virtue of domination, and the simplest way to challenge such power is to challenge its source: knowledge. This kind of challenge to power thus leads to a challenge to the value of the occupational information: all things appear simple to the simple.

We find this expressed again and again. As long as we consider man solely as a tool maker and therefore as a commodity producer, and as long as we are content to balance this occupational aspect of his activities with empty phrases such as "the quality of life", "human dignity", etc., power will continue to be based upon an occupational hierarchy which depends in turn upon the degree of abstraction in occupational information.

We should once again recall that power depends primarily on its functional indispensability to the whole, i.e. to the human grouping under consideration. Any individual or group of individuals which is not indispensable to the structure of a set has no reason to hold "power", since the set can be sure of performing its function without them. We must now see whether this indispensability depends on the abstractness of the occupational information possessed by an individual or group of individuals, by what we have defined as a "functional class". Having done that, we shall then have to determine what kind of power is at issue, political power or hierarchical power.

Gérard Mendel[41] gives an example concerning the relationship between teachers and those taught. Clearly, it is not occupational information that gives or ought to give, to those who are taught, any power as a functional class. But they are just as indispensable to the collectivity as the teachers, and by virtue of this they ought to hold or to win power. It is likewise clear that the amount of information accumulated by a teacher in his discipline is very often more important than that of the bureaucrats, but he is dependent on their power.

So we see that in contradiction to the general formula which I have developed so far, political power does not always depend on occupational power, nor on the latter's degree of abstraction. Thus although we can say that the indispensability of a social group in societies as they are today often depends on the degree of abstraction in the technical knowledge of the individuals composing that group (an idea that underlies the increasing power of the technocrats), there have in spite of this been strikes to show us that there is a kind of indispensability which exists apart from any degree of abstraction in the technical knowledge of the people on strike.

We need to note that the strength of such strikes also depends on the number of strikers, and that this number is generally inversely proportional to the degree of abstraction and specialisation in their occupational knowledge. But when the dustmen in New York, who are right at the bottom of the hierarchical ladder in terms of the degree of abstraction in their occupational information, went on strike, the whole life of the city became impossible. The strike allows us to tackle some interesting ideas. By making it clear that a particular human group is indispensable to the life of an enterprise, an industry, a city or a whole nation, the strike allows a certain kind of power to be exercised in an exceptional manner.

By means of strikes and the unions, the dominated have won certain advantages, particularly economic advantages, that the dominant would never spontaneously have given them. But let us repeat that this power is chiefly economic, and hardly political at all. It is not the decision-making power that is at stake either in the enterprises, or in the industries or at the national level. In other words, the advantages have still been won essentially at the thermodynamic level, and hardly at all at the informational level. And one can guess why. It is essentially because those who are dominated are dominated only because of their informational deprivation. Even when there is a strike by highly specialised groups (for

120

example, a pilots' strike), the information which they express and which they manipulate in order to press their demands about money or working conditions is not generalised information; it does not concern the role of their functional class in the enterprise, nor that of the enterprise within the industry, nor that of the industry within the national set, nor the general finality of the latter on the planet. Their action is based upon corporatism. Corporatism is extended only by opening up in a horizontal direction, not in a vertical one. One may even have a strike of all the pilots of all the international airlines, but one rarely sees a strike affecting all the hierarchical levels within a single enterprise, industry or nation, for then the finality of the largest set would be called in question. Whatever the system may be, this rarely happens. The probable reason for this is that it could only happen if each level in the occupational hierarchies were ready to accept the loss of its position of domination. Now whatever the economic dissatisfactions may be, they are relatively trivial in comparison with the satisfaction obtained from domination: the gratification which stems from multi-level hierarchical power compensates for the lack of gratification which stems from having no political power. (This is one of the reasons why strikes are generally embarked upon by the most underprivileged social strata, because it is they who are stuck at the lowest level of abstraction in occupational information; the thermodynamic factor remains the preponderant one in their labour, and their hierarchical gratification is the lowest.)

This conclusion would seem to contradict the one we reached earlier, namely that the occupational power which is linked to specialised information extends to and is fused with political power. We shall now see why these two statements do not contradict each other but, on the contrary, reinforce each other.

Democracy seems necessarily to generalise political power, since the vote of a managing director counts for no more in the ballot box than that of a machine hand.

However, it must first of all be noted that because of the growing importance of specialised information in comparison with that of human thermodynamics, an increasing number of people are participating to a greater or lesser degree in hierarchical satisfaction, and so find themselves less tempted to unite with the underprivileged, since they are unconscious of their own total lack of political power and do not even try to claim it. "I don't involve myself in politics, you know." People say that to you as if they were telling you that they do not suffer from some infectious disease, without realising that it is roughly the same as saying that one is blind, deaf, dumb and impotent. It is true that to be involved in politics in the way that people generally are involved, moreover in the way that most politicians are involved, is roughly the same. But the important point is that the idea of information and of structure allows us now to understand that the managing director and the machine hand, as social elements, both have relatively little importance in comparison with the relations and connections which they have in the social whole. In other words, what is important is the social structure, the whole set of relations between the elements in the social set and not the absolute number of managing directors or machine hands. One can imagine quite considerable changes in the numbers of these elements, without the general structure of the whole being greatly changed because of them. In a hierarchical structure it doesn't basically make a great difference whether the dominant elements are capitalist, bourgeois, technocratic or bureaucratic, or whether there are fewer machine hands and more technicians. The labels attached to the elements may change, but the hierarchical vertical structure stays the same.

Under these conditions democracy is obviously a game for suckers, once political power has been delegated within a hierarchical structure. This is why, in spite of the disappearance of private ownership of the means of production, the political power of the individual has not grown (to say the least) in the present-day systems described as socialist.

It follows that what a voting paper expresses is not real political power informed in a generalised way, nor power based on knowledge and indispensability, but rather the acceptance or rejection of a hierarchical system which extends an occupational hierarchy to the political arena, according as the individual feels or does not feel sufficiently gratified by his hierarchical occupational position.

The vote is not about a fundamental reconsideration of the overall finality of the State, in spite of the stereotyped phrases that are hurled against "capitalist profit" without putting at issue the whole question of expansion. It is not about "the quality of life", if hierarchies and industrial society are not called in question. It is not about small traders, small farmers, craftsmen and so forth unless the indispensability of the functional classes is called in question. For the real exercise of political power demands the

122

possession of information about problems posed in relation to the general structures of the national set within the international sets; moreover this means being informed in a "non-orientated" way, which I am going to define later on as "generalised information". The "new societies" have never been able to conceive the elements which this generalised information needs in order to establish itself: the time necessary for each individual, the polymorphism, the scientific structures (socio-biological ones in particular). These new societies are simply societies for increasing the economic satisfaction which is linked with growth; satisfaction of the need for political power, which cannot be achieved without generalised information, has never been envisaged. The "social malaise" springs up here too. Since the system is founded upon a hierarchy of occupational power ranging over a very wide spectrum, each individual finds an "inferior" from whom he obtains gratification through paternalistic domination, and a "superior" who prevents him from obtaining gratification and causes him to be alienated. But he also finds an institution which gives him security for the future and is maintained for the satisfaction of his fundamental needs. The individual is neither happy nor unhappy: he is conditioned by the mass media in such a way that his motivations are entirely directed towards commodity consumption and social advancement, which perpetuate the hierarchies of value and income since these are entirely organised by commodity production. Democracy in the "free countries" (an expression intended of course to influence public opinion) shows that most individuals, in total ignorance of these fundamental problems, vote for those who promise them an easier life or those who promise to protect what they already have. The result of the vote tallies with people's consciousness of their own gratification in a given system, according to the extent to which they are satisfied with their domination-status. And when they are not satisfied the vote goes against one system and for another, which never fundamentally calls in question the occupational hierarchies of domination or the problem of growth. One votes for a system that will wholly reproduce the occupational hierarchies which are a fundamental source of alienation. Only the label changes.

It should not be thought that the dominant possess more real political power than that which is necessary to maintain their domination. Of course they possess "the" political power, in the sense that what is conventionally called information and the means of disseminating it, the mass media, are completely at their disposal. They are thus able to direct public opinion and people's needs, and by virtue of universal suffrage they give the impression that democracy really exists. Of course it is they who direct the big firms and the banks, and they enjoy the support of politicians who give legal expression to their decisions. But here again it is not the "capitalists" that matter but the "structure" in which they act. If these capitalists, who act only to preserve their hierarchical domination, were to disappear, so long as the hierarchical structure remained they would be replaced by technocrats and bureaucrats, whose motivation is identical even if the means used are not always identical. Profit is only one means of securing domination; the police, confinement in psychiatric hospitals or in concentration camps are others, as too are spying, phone-tapping and hidden microphones. But the most effective and widely used means is the automatisation of thought and the creation of conditioned reflexes and value judgements. Education and the mass media, in the hands of the power structure (that is to say the hierarchical system), have no other function.

This shows that the rules for gaining domination are institutionalised. This institutionalisation certainly constitutes the occupational hierarchical structure which permits the acquisition of political power; but it is a false political power, since its sole raison d'être is to maintain the domination of the dominant and the submission of the dominated in a process of commodity production.

In my opinion, the fundamental obstacle to the realisation of a socialist society is the existence of hierarchies, the distribution of economic and political power according to a scale of values which depends on productivity and is measured in terms of commodity production. When a social structure is not involved directly in the system of production, it is involved in protecting its hierarchies: the army, the legal system, the police, the bureaucracy, art and what we conventionally call culture.

Where are we to locate the class of "workers" and their class interests? High-grade technicians or skilled workers probably feel that they belong to the proletariat, the class of "workers", to the extent that they fail to experience the satisfactions of hierarchical domination. On the other hand, in the working class you can find the perfect bourgeois, who is happy to be so even though he is exploited and robbed of his surplus value. Similarly, in the bourgeoise there are people who are genuine proletarians and proud of it, although they fully enjoy their economic and political power and accept it as being quite fair, since they do not question the existence of hierarchical power but only how it is distributed. Up to the present time the idea of class has been based solely on the possession or non-possession of economic and political power. This economic and political power is itself based upon a hierarchical system which is a function of occupational information. As long as the parties of the "left" do not call in question these very foundations of the hierarchical system, the class struggle will always be born again out of its own ashes, since the system that gives birth to it will not have been abolished.

In this sense, there exist the dominant and the dominated, whom we can certainly call bourgeois and proletarians, if the occasion arises. We can apply to each grouping the expression "social class". We can certainly admit that the effort of the dominant to maintain their domination and that of the dominated to achieve it constitutes the "class struggle". But equally certainly, it seems to me that we are limiting ourselves to mere revolutionary phraseology if we do not introduce into this framework the set of ideas that we have broached regarding information and thermodynamics, occupational hierarchies and political power. These ideas make it much more difficult to define the boundaries of the "classical" social classes. We now know that these classes are characterised by the relation between abstractness of information and mechanical labour in the activity of individuals. The higher this ratio, the higher the class. It is this relation that furnishes the "power" to act, since the better informed the action is, the more effective it is. We have seen that this power is to be found in the occupational hierarchies, and that it becomes a political power by virtue of the fact that "politics" has so far never done more than maintain the power of the dominant (conservatism) or seek to take their power away (progressive politics, revolutionary politics, leftist politics) while remaining within the current framework of economic growth. The very *structure* of society, the hierarchical structure, has never been made the basic point of the dispute. The outcome of this is simply the replacement of certain *elements* (the capitalists) by others (the technocrats or bureaucrats), and the problem of finding out what are the bases of the hierarchies, their meaning, has never been posed. This would lead on to questions about the global finality of the human species, and this would then have had to become the matter at issue. Is man programmed by evolution to be essentially a commodity producer?

Finally we need to clear up a misunderstanding about the idea of power. The person whom we conventionally call an intellectual, particularly one who is specialised in a certain technique, has a certain "power". If, for example, he shows himself to be an effective propagandist for the value judgements that constitute the armature of the society in which he lives, he will as a result be rewarded. A place to work, access to the means of disseminating his clichés, "honours" and academic rewards will be bestowed upon him for playing the role of the true humanist, who displays such an elevated mind. Of course, we know very well that minds only get "elevated" within the dominant ideology, the ideology that ensures the solidity of the existing hierarchical structures. By means of a smooth shift of position (for which there is no justification whatsoever but which fits the system well enough), the specialised credit which he has accumulated in his discipline is used to prove to the public the validity of his conformist value judgements on problems of a quite general character. Besides, in what does the "power" of an individual reside when he has never done anything but reproduce? Power must in fact be accompanied by access to the means of coercion. Real power is accompanied by the means of ensuring that what one says is listened to, whether this comes in the form of money, access to the press, police assistance or the removal of competitors and rivals from the rungs of the hierarchical ladder.

One can hardly then accuse a person who puts out new ideas of possessing or trying to gain power, if he has no coercive means of getting his ideas adopted other than the general agreement which is accorded to every discovery when it has been confirmed by many experiments carried out by other people. Can we say that when Galileo said in stage whispers "and yet it turns", he possessed power? Was it not the tribunal, which had just forced him under threat to express an opinion conforming to the prejudices of the times, that held the power in its hands? The power accorded to the opinion of a specialist in his discipline is surely confined to the expression of this opinion within the ideological conformism of the society in which he lives. Can a creative person really enjoy power? Transforming social or conceptual structures will never enable him to enjoy the means of coercion which the individual who conforms to the system can enjoy. He belongs to a hierarchical system, he abides by it, and one of his rewards is to decide who will succeed him in the evolution of the hierarchy within his discipline, basing his judgement on how well they conform to the system. The creative person does nothing but supply new information: he has no means of coercion to make people accept it. It is precisely for this reason

126

that new information is usually so slow to be generalised, and so difficult to get people to accept. This is also why breakthroughs into new fields of information are also so rare, for they bring little by way of reward in a solidly structured hierarchical system. It is true that the biologist more and more often meets people, particularly amongst protagonists of what are called the human sciences, who are in revolt against the kind of "power" which the biologist would like to acquire and impose, as well as people who are in revolt against the so-called "imperialism" of biology. We are dealing here with people whose system of knowledge is closed; they refuse to be included in a wider system in which they could play the role allotted to their discipline as a sub-set of human knowledge, and thus are afraid to see their domination wane before a new set of knowledge. Instead of making the necessary effort to be informed, to go outside their closed system and establish interdisciplinary relations, they attribute to a new set of knowledge — which wants only the pleasure of existing — a style of behaviour belonging to their own discipline, which they practise with the object of obtaining hierarchical satisfaction. In fact it is not "biology" which displays a kind of imperialism. What annoys them is the fact that biology, which is an interdisciplinary discipline by definition, can't do without biologists, whereas they would pour boiling oil on anyone who tried to invade the citadel of their own discipline without the due protocol.

The aggressive attitudes which are beginning to appear against biology are born of the distress which it provokes in some people because it is the vehicle of an unknown complexity; they can be explained also by the fact that its field of action ultimately relates to the conscious and unconscious behaviour of human beings. I have already pointed out that there has so far been nothing in common between physics, which is rigorous and mathematical, and speech. This has left the field open for kinds of discourse in which the dominant unconscious finds expression. Now we are beginning to see that the dice used by the human sciences have been loaded, that the rigour of the physical sciences needs to be gradually extended to the sciences that deal with the living world. All the talk of transcendence and essence, behind which it was so easy to hide our ignorance, begins to lose its spellbinding power.

Having talked a lot about power, we have reached the point where its very existence has become the point of the debate, rather than the rigidity of a particular system. This once again presents a confusion between information, finality, function and structure. The function of a schoolmaster is to transmit certain information which prevents the student having to travel the hard road of human knowledge from the prehistorical epoch all over again, on his own. It is not his function to impose this information, and so in order to do so he has to use means of coercion. If this happens, the finality of the student is no longer his own, and in the structured set which represents a class not each of the elements is pursuing the same finality. What must be done, then, is to try to give one and the same finality to the basic motivations of all the elements. But the motivation of the teacher is not so much to transmit information as to fall in with a certain imposed programme and, by going along with this programme, to aid his own hierarchical promotion. The motivation of the student is that the socio-economic structure expects of him, namely to become a participant as quickly and as effective as possible in the production process, by acquiring an occupational training. On the other hand the student is also looking for paternal reassurance from the person who knows, the teacher. And the person who knows, the teacher, may perhaps be sufficiently gratified by the childlike submission of the student. In other words, the "function" of the student is essentially to receive the information which corresponds to his motivation, but this is usually transformed into the soothing alienation of a "power" which has no *raison d'être*. Similarly, the "function" of the teacher is to transmit information in response to the motivation of the students, but this function is transformed into a paternalism that gratifies the desire for domination and "power", which has no functional justification. What applies to teaching applies to the set of society, a closed structure which is itself composed of individual closed structures whose motivations differ.

9

Open and closed systems in sociology and economics

It is impossible to establish a generalised political power while maintaining occupational hierarchies as the criterion of power. On the contrary: power, which up to the present time has not been really political but simply the power to press demands or even just to eat, arises when a functional class (by means of a strike, for example) makes the social set in which it is situated feel its indispensability. Clearly, then, a generalised political power must rely upon the indispensability of the functional classes. This leads us towards a more precise formulation of how the concept of an open or closed system can be applied to economic and social structures.

Above (page 24) I developed the idea of "information-structure", showing that all levels of organisation are closed systems, and that their opening only becomes possible as a result of their inclusion in a larger set, at a higher level of organisation, on condition that circulating information permits this integration by transforming the closed system (acting as a regulator) into a servomechanism. I made the point that such a structure is closed at the level of the individual organism and that if we want to effect its opening, we must include it in social groups which unfortunately themselves also represent closed structures and are therefore antagonistic. On the other hand, we know the following things about every closed information-structure at a given level of organisation.

(1) It can only have one finality, namely the preservation of this information-structure. All its global or partial functions are only *means* of attaining this goal. A heart, for example, is indeed a biological pump whose function consists in mobilising the stock of blood. But one should not make the mistake of believing that this is its finality. This is simply its means of maintaining its own organised structure and thereby participating in the maintenance of the structure of the organism as a whole. If this latter ceases, then the structure and function of the heart also cease, just as they cease together with those of the whole organism if the cardiac function

is no longer assured. A heart continues to beat in a preservative liquid, but only because the experimenter who has isolated it and separated it from the information and structure in which it is ordinarily to be found, who in other words has transformed the servomechanism into a regulator, meets its energy requirements. He observes the reaction of the regulator by adding, to the liquid in which it is preserved, molecules of the "circulating information" which under normal conditions modulates its functions: hormones, the chemical intermediaries of the nervous system, mineral ions, osmotic or hydrodynamic pressure, the pH value, the concentration in different metabolites, carbon dioxide, organic acids, various chemical products, etc.

(2) We know also that this information-structure, a closed system, is in fact an open system at the thermodynamic or energy level and at the level of circulating information. With regard to the second of these, certain ideas need to be recalled. We know that an organism picks up information in its environment and transforms this information through its sensory systems into coded signals; these are sent through all the levels of organisation that play a part in the constitution of its information-structure. Information issues from the latter only as the movement, action and behaviour which it directs. But it is not only these things which direct behaviour: a part is also played by all the internal circulating information resulting from the condition of well-being which exists in each element of the organism, and which is necessary for the effective accomplishment of its function. And so the circulating information which results from the opening of the system on to the environment, closes once again over the latter in order to transform it in accordance with the indications supplied by internal signals (see figure 8). The finality of these signals is, as we have seen, to maintain the information structure of the organism as a whole, a system which will remain closed if it is not opened by inclusion in a larger set, the social set.

So we can see that the now commonplace idea that living systems are open systems is incomplete if no distinction is made between different types of opening and different types of information, and if it is not in addition noted that these openings have only one end, which is to maintain the closure of the information-structure at a certain level of organisation, i.e. that of the individual. This is the sole end achieved by the continuous energy-opening which transforms the energy taken from the environment. It is also the end achieved by the informational opening which enables behaviour

130

to act upon the surroundings. The *sociological* problem, therefore, is to define the social information-structure which has enabled the organisation of the living world to reach the stage of complexity where the human species has arrived. In other words, the problem is to define the information-structure which permits the opening of this closed system — the individual — on to a more or less complex set of individuals achieving the level of organisation above. The *economic* problem is to determine precisely how the mass and energy taken from the environment will have to circulate in such a structure in order to ensure the maintenance of its global organisation, while at the same time ensuring the maintenance of all the indvidual elements which constitute it.

The planetary society

As far as the sociological definition is concerned, it would seem obvious that the finality of the set must also be that of each of the elements which constitute it. This finality is that of maintaining the structure. If the information-structure of the social organism is a hierarchical one, it is obviously difficult to convince each element in this progression of hierarchical levels that he must work in order to maintain his own alienation. Maintenance of the structure is achieved rather by giving him hope that he can rise in the hierarchical system, by preaching social advancement to him, by telling him that if he submits to the system he will one day be able to alienate the alienated people among whom he belongs at the moment. This brings us back to the advantage of having the kind of hierarchical system which is staggered at many levels, to the ensuing malaise and to the rarity of explosive crises (see page 98 above). It also brings us back to the fact that hierarchical systems today are based on the degree of abstraction in purely occupational information necessary for the commodity production by which domination is established. This is also dealt with by making each element in the system believe that the way things work out from the very start depends upon the innate structure of the individual organism, which itself is an accident of the genetic combinatorial. Some people are gifted, others are not; some people ascend the hierarchy by talent alone, and honours, power (in fact, only the power of consumption) and wealth are due to them because they are more "intelligent" than the rest. The word "intelligence" is used with extraordinary looseness; it is the empty notion which governs our contemporary world. Success (in the hierarchical system, of course, for it is this alone that counts) depends on intelligence. When I spoke of the nervous system, I noted in passing that I have never encountered a "centre of intelligence" there. We can act upon mechanisms which, though complex, seen today clearly to exist: *basic motivations, acquired motivations* (acquired by the *memory,* which calls these motivations into play), *concentration, waking, sleep, aggression, affects* (that is to say, all our "emotions"), and *imagination.* But I have never come across any definition or mechanisms for intelligence. This is because "intelligence" — an all-purpose expression which is the open-sesame for social success — does not ensue solely from the information-structure given at the outset, from the fortuitous coming together of a "gifted" spermatozoon and a "gifted" egg. Rather, it ensues from the development of an individual lambda information-structure in a particular environmental situation. To the extent that this development conforms to the criteria (the IQ scale) of the social set, to its finality and its hierarchies, the nervous system expressing it will be called intelligent or otherwise. This does not mean, things being as they are today, that there are not some individuals who are better able than others to "comprehend" (in the full etymological sense of the word) the information which they receive and the structures in which they are enclosed. These differences arise from interactions between the nervous structure and the environmental niche which are so complex that we cannot accurately analyse their innumerable historical factors. This does not mean that these individuals will be considered by their contemporaries as more "intelligent". It means simply that they will be different. And to admit difference — that is to say the unknown, the generator of distress — is the most difficult thing, as we already know. We shall return to these problems when we discuss the question of creativity.

So, since the finality of the set can only be the maintenance of its structure, the means which it uses for this purpose would seem to present a problem. Perhaps the human species, like all preceding species, is biologically linked to a structure of hierarchical domination. If that is so, all efforts of the imagination to find an alternative structure are vain and utopian. However, our field of consciousness cannot help being impressed by the existence of certain organisms, in all living organisms in the biosphere, in which a collectivity of cells lives — that is to say, maintains its structure — without a hierarchy of power or value. The purpose of this book throughout has been to ask why the hierarchies exist, and we have reached the conclusion that the reason for it is the closure of the information-structure at the level of organisation of the individual. As soon as one proceeds to the level of organisation of the social group, domination appears, hierarchical structures form and the leader emerges, with his little clique of privileged people and a mass of slaves scattered all the way down the hierarchical ladder. We know that this stems from the inadequate circulation of mass and energy on the one hand, and of circulating generalised information on the other, within the social body. Now suppose this problem has been solved. Imagine a social group without hierarchies, and with equal distribution of energy and information. This group structure will be closed, and will enter into conflict with the information-structures of groups of equal importance which surround or encompass it: this leads us to the notion that a socialist structure, or a self-managing or non-hierarchical structure, can only be universal. If from the point of view of the information-structure it is merely a closed sub-set, it will try to get energy and information for itself with the sole purpose of maintaining its own structure, at the expense of those around it; otherwise it will be absorbed by them and disappear. Unless, that is, it can live by contenting itself with the energy it takes from its own ecological niche, from its own territory; and unless it closes its frontiers and suppresses the exchange of energy and information by means of an "iron curtain" which neighbouring structures with expansionist desires cannot break down. If this structure does not succeed in maintaining its isolation, it will be in a very unfavourable position in competitive struggle with the other structures in its environment, for these are hierarchical and organised around commodity production and economic expansion, and are thus only able to find the energy necessary for their survival by going out of their own ecological niche and taking it from elsewhere, by means of domination. Domination springs from the use which they can make of specialised occupational information. It is by this same mechanism that individual domination is established in the hierarchical ladder of the group. It is occupational information of a technical variety that enables a human group to make more dangerous, more effective and more numerous weapons with which to go outside its own ecological niche and, by means of domination and fear, appropriate energy and mass (raw materials) from ecological niches where other, less technically developed social structures are situated. It thus acquires wealth without, however, solving the internal problems that ensue from the hierarchical structures; these are its own specific problems, and they often take racial forms. If another structure reaches a similar stage of technical development with a different hierarchical system, the antagonism can temporarily be resolved by means of a type of equilibrium, the "balance of terror".

Some day we must bring about such an information-structure for the set of human beings on this planet, suppressing international hierarchies and domination, if we wish to avoid the disappearance of the species. But this presupposes a profound change in the behaviour of the individual, since each individual will in that case have to act for himself as if he were acting for the whole species. It is not that he will have to act "freely"; but the structure in which he is included must motivate him in such a way that he cannot have any other attitude, just as at present the hierarchical structure forbids him to act otherwise than to maintain this structure. His acts of gratification must go towards the maintenance of the social structure of the species. Since this structure will no longer be founded upon the domination of groups, we may moreover note that there will no longer be the sacrifice of life in periodic holocausts called wars. Because competitive expansion will no longer be the sole finality of human groups, the individual will no longer be called upon to work without joy. The time spent on work without joy will diminish in proportion to the constantly improving use of technical information in making and controlling machines; correspondingly, needs of socio-cultural origin, created in order to ensure economic and hence hierarchical domination, will not grow. Improvement in the conditions of life will simply be a consequence of developments in knowledge. The growth of needs will no longer be regarded as a finality, but as a necessary by-product of increased knowledge. Education itself, which has never been looked upon as anything but the means of moulding individuals for commodity production or for the permanence of the hierarchical structures, will take on a different meaning and a different content. We already have an example of such a chain of events. No one can dispute the fact that the great contemporary discoveries in the field of biology, for example, which have made possible the almost complete disappearance of great world-wide epidemics, have not generally speaking been *motivated* by the need for economic expansion and pursuit of hierarchical domination, even though they have been made within such a system. Furthermore, the same might be said of all the great fundamental discoveries that have turned today's world upside down. But in the immediate present these discoveries have been exploited by human groups with the

134

object of domination in either the commercial, the economic or the military sphere. Each sphere complements the others: the pharamaceutical industry and the war industry are both motivated simply by domination, by virtue of the fact that they are part of a national or monopolistic information-structure.

In order to motivate the individual to seek generalised knowledge rather than the technical knowledge necessary to rise in the hierarchy, it obviously would be helpful if he were to have some gratification. To put it conversely, he has to establish a relationship between the fact that technical advance ensures his individual gratification and the fact that it is punitive, coercive and dangerous for the species and thus (by way of "fall-out") for himself and his descendants. All the warnings put out by the ecologists, the Club of Rome, etc., do no more than assert one fact: if we extrapolate from future trends, the human species has set course for destroying the biosphere itself. With this as our starting point, we have tried to discover its mechanism; other groups have failed to do this, because they have not introduced into their research the biology of human behaviour in the social situation. The energy crisis, of which the species has become conscious at an even more recent date, certainly constitutes an additional factor in this range of factors representing yet another form of the pressure of necessity to which all evolution has been subject. Perhaps the human species will be fortunate enough to interpret the message of all these events taken together in good time, and to transform its group structure and finality in such a way as to escape disaster.

In this planetary social body, which is utopian but sooner or later necessary, the hierarchical structure founded upon occupational information must disappear, since the finality of the species will be unable to tolerate the domination of social groups acting in their own interests. This greater set will have to find its finality in itself, and this finality is hardly likely to be commodity production, for the biosphere is limited in space and time. The sources of energy are also limited, even with the taming of the atom and the improved utilisation of solar energy. Supposing even that we were to find means of guaranteeing energy resources that are for practical purposes limitless (which basically is not beyond the realms of imagination, for we are actually beginning to understand what this means), then if the information-structure of the integrated organism of the species were no longer geared to the pursuit of internal domination, it is probable that the finality of this set and of each individual in it would still be knowledge. One of the basic

135

motivations certainly will not disappear, namely humanity's response to the anxiety of existence, or in other words the anxiety of death. This seizes every human being by the throat from the moment of his or her consciousness of existence and does not quit until the moment of death, but it is persistently covered up by contemporary societies, for it gets in the way of their finality, which is commodity production. One even wonders whether this is not an important factor in the establishment of hierarchies. When someone is preoccupied with his social advancement he is less preoccupied with the meaning of his own existence, and is therefore a more effective agent in the process of production. One wonders whether the person who succeeds best in such a process, the person for whom hierarchical promotion is most assured, is not also the least human, the least conscious (I almost said the least "intelligent"), the most blind, the most automatised, the most satisfied, the most gratified by his domination, the least uneasy: the truly "happy imbecile". And if I have a liking for the youth of today — the hippies, the "antis", the rebels, the aggressive down-and-outs and the social failures — it is because very often these people have in some obscure way seen this anxiety of human existence and have not been able to content themselves with the social advancement that is offered in exchange for submission to the system. I am also tempted to find fault with the revolutionary ideologies for trying to turn this anxiety to their advantage, that is to say to the advantage of groups seeking domination. Of course, it is only by seizing domination from those who now possess it that another system can be put in its place. But by providing rigidly fixed grids to channel the action necessary for the appeasement of anxiety, one also rules out imaginative research and the understanding of the mechanisms that establish structures of alienation in society today. All that becomes possible is the replacement of one system of domination by another.

Let us now move down from the social set at the planetary level to its national sub-sets. The geographical criteria by which national areas are demarcated will no longer perhaps have much meaning in the near future, and will probably be replaced in importance by linguistic and economic considerations. As long as this is the situation at the national level of integration, language will without doubt be a considerable obstacle to the realisation of a planetary society. It is hard enough to make oneself understood even when using the same language. A father and a son don't understand one another even though they speak the same language, for their experience of words is not the same, and their conceptual reactions and value judgements (linked to vague words like freedom, equality, fairness, honour, law, duty, discipline, etc.) are necessarily different.

This, together with economic dependence, is a fundamental cause of conflict between generations. It is even more difficult to make oneself understood when two different languages are being spoken, transmitting what we conventionally call two different cultures, that is to say transmitting the set of value judgements and prejudices shaped over centuries by a particular ecological niche. So it seems that generalised bilingualism will be necessary for a long time, with simultaneous use of both an international language on the planetary scale and a local language belonging to the sub-set; this will allow people to preserve the experience accumulated over centuries by sub-sets of the human race which have grown up in particular ecological niches belonging to them. In fact the recognition of regional collectivity is like the recognition of the individual, of differences in other people and the acceptance of these differences. The suppression of these is an attempt to suppress the anxiety which they nourish because they make us aware of gaps in our information and understanding; the recognition of particularities and differences, on the other hand, is a fundamental basis for the inclusion of a sub-set in a larger set. It is the fundamental basis for the opening of the information-structure. It shows respect for a law of biological evolution which resides in the genetic and informational combinatorial. (Imperialism and racism have been established in ignorance of this law.)

One can understand the extent to which, from this stage of development onwards, economics is influenced by sociology and vice-versa. The interconnections between them are dangerous only if economics is, as now, the criterion of domination; certain regions of France such as Brittany and Vendée have experienced something of this, particularly in the centralised and rigid administrative structure which is rampant in France. All the problems of elaborating regional structures can be summed up as the problem of demarcating the sub-sets in such a way that the different sub-structures can coexist harmoniously in them: linguistic, historical, geo-climatic, psycho-sociological and economic sub-structures, none of which should be favoured at the expense of the others. The current tendency to favour the economic sub-structure results from the view that man is solely a machine, to be engaged in production within the particular ecological framework in which he finds himself. In fact this is only partly true. The sense of well-being that one gets from living in a particular region does not always spring from its economic development, but conversely the latter may be important for the economic development of the national set. It may also be the case that if in the near future the human species abandons, out of necessity, its present finality of expansion and the domination of social groups, then the demarcation of human society into economic regions will perhaps take on a new character. The demarcation of regional sub-sets would not have to be imposed from the top, from some "central office of administration", but would be raised first of all at the base, on the basis of mutual attractions and adjustments. Ranging from individuals to local communities, these would weave a complex net of varied interests, antagonisms or co-operative associations motivated sometimes by automatisms inherited from way back, against which it would then be useful to set more global perspectives derived from integrating schemes and imaginative complexes. The broad dissemination of information, the exposure and denunciation of narrowly-conceived selfish standpoints, of the crude motivations which guide them at all levels, of their paternalism, their concealed dirigisme and their search for domination: all this would be the necessary condition for any temporary attempt to establish an administrative sociological framework, a territorial ecological niche.

These ideas may seem exasperatingly commonplace. But the point is to build a socio-economic information-structure, starting from the smallest possible sub-sets, the individual and the family, and finishing up with the largest national or international and planetary ones. How can we do this without disseminating generalised information, unburdened as far as possible of value judgements, and generalised to the set of individual elements first of all by teaching and thereafter by various means of dissemination? In this case is it not necessary to choose a finality for the greatest possible set to which the finalities of the sub-sets will for their part contribute? The finality of growth has so far meant that education, "culture", the army, the police force and the magistrature are necessarily orientated towards occupational training, the protection of the hierarchies which spring from it and from private property, and the global internal and external defence of the system as a whole. Let us suppose, in contrast to this, that the primary finality becomes education. When this is not directed solely towards production, growth becomes only a secondary consequence of education instead of being its end. The rigidity and exclusivity of hierarchical educational systems will certainly then tend to grow weaker. And since the essential part of what must be defended will be structural rather than material, "the nuclear arm" may, like the army and the police, become less indispensable. Obviously all this is still a fairly vain hope. But it is perfectly realisable: it depends on the formation of teachers who can to some extent be freed from their paternalism, their value judgements and certainties, and from all the closed systems by which they are encumbered. In the space of a hundred years, without the help of modern audio-visual aids, republican teachers have managed to generalise bourgeois education and its social prejudices. It ought to be possible today, with a little enthusiasm, to generalise more rapidly a form of education which is relativistic, dynamic and evolutionist, and to allow children's brains to remain open systems which shun slogans and cut-and-dried ideas like the plague.

The problem posed by the discovery of a new kind of information which structures human societies is therefore mainly one of finding a new finality for the human race. Economic expansion is not, one imagines, a phenomenon that can go on forever within a biosphere that is limited in space and time, even if the day of reckoning is deferred by discovery of new sources of energy or by the control of pollution and population.

From then on we may have a new epoch for the human species. The occupational hierarchies of value will be less menacing because they will be less finalised, and other social structures can come into being. We shall try later on to imagine what these social structures might be like. Let us for the time being simply look at the possibility of changing them, reluctantly accepting the fact that their finality will remain as it is for a certain period of time.

The planetary economy

Our approach compels us to tackle the *economic* aspect of the problem. I have no intention here of invading the territory of highly qualified specialists whose techniques are foreign to me. I only want once again to describe how a biologist situates economic phenomena within living sets.

I noted at the beginning (see page 20) that all living forms in the biosphere can be considered the result of a specific "giving of form" to matter by means of the luminous energy distributed over our globe by solar photons. It is this energy which has enabled the information-structure of living systems of each species to appear, and to evolve from the most primitive forms into the most complex. It is this energy which, by temporarily maintaining the information-structure of each individual in the form of nourishment, has permitted the survival of species throughout generations. This nourishment itself is simply the energy from solar photons, transformed into chemical energy.

Besides yielding these consequences at the level of the organisation of species and individuals, solar energy is also the source of the mechanical energy which each individual releases and which is expressed in behaviour, i.e. in his action on the surroundings. Nourishment therefore guarantees both "labour power" and the conservation of the living structure which is labour power's material support.

In the human being this material support (whose structure is maintained by solar energy transformed into chemical energy in the form of food) is, for the first time in the evolution of species, itself a source of information, thanks to a perfected nervous system whose operational mechanisms I have already described. By means of this information, human beings transform matter and use energy. They do this at first to protect their physical existence within their environmental niche and then, increasingly as the centuries pass, to establish and maintain their hierarchies of domination.

An event of some importance has occurred quite recently in human history: because of this processing of information which is unique to the human species, man has discovered the means of not being entirely dependent on solar energy. He has discovered the means of using atomic energy. Up to then his energy resources — coal, oil, fossil fuel and hydroelectric energy — were derivatives of solar energy. For the first time human ingenuity has discovered a new source of energy, and we don't yet understand its importance very well, nor the impositions it may make upon us.

A vegetable uses solar photon energy for the sole end of maintaining its structure. An animal also does this, but to attain this end it is not restricted to using the nutritive elements biochemically; it can go out and fetch them in a movement which is autonomous in relation to the environment. The environment then confers on it the ability to transform chemical energy into mechanical energy. Finally, the human being does what the animal does; but he can also use the mechanical energy he releases to bear information, because he is able to process the information which is stored in his

140

memory, which comes to him from the environment and moulds his imagination.

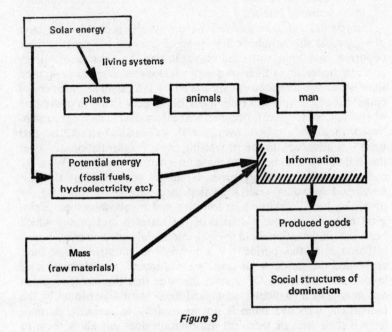

Figure 9

Man is thus dependent (see figure 9):
(1) On matter (raw materials) which he finds within his ecological niche or which he may be tempted to go and seek outside it.
(2) On matter once again, but matter which is already transformed by solar energy into vegetable or animal foodstuffs. In this case, moreover, man's inventive ingenuity has improved on the use of solar energy with his discovery and constant development of cultivation and cattle-raising. These were only possible because of the discovery of certain uses of raw materials, such as chemical fertilisers, and the discovery of other things such as metals for the construction of machines, which permit industrialised cultivation as opposed to artisan cultivation. But these machines need:
(3) a source of energy, generally fossil fuel.

It is then essentially because he uses
(4) increasingly abstract information that this progress is possible. Thanks to this, the populations of the industrialised countries have entered an era of abundant food and have eliminated the spectre of famine.

But clearly, industrial development would probably not have developed at the extraordinary speed of the last 100 years if these countries had been restricted to exploiting the sources of energy and raw materials in their own ecological niche. They have secured their economic development by means of the organised robbery of countries which were not technically developed. Being in possession of specialised, technical, occupational information, they have established their domination over social information-structures that were less advanced in the processing of such information and over the individuals who belong to these information-structures. What we currently call the energy crisis is simply the result of the less developed countries' realisation that they were being exploited, by providing local workforces at low cost and by allowing their ecological niche to become a source of the material and energy which bear the information held by the industrialised countries.

Throughout this period the industrialised countries alone have consumed the products of this "well-informed" transformation of labour and materials. Of course, the idea that the material goods and energy of a particular ecological niche should be owned by the individuals who live there is not necessarily self-evident, particularly if their lack of technical information does not allow them to take it up. Nevertheless, the idea of appropriation by a social group ensues from the closure of the information-structure which the group represents. If the material goods and energy of this planet could be reckoned up and divided out in such a way that they belonged to no one in particular, to no human group, but were recognised as belonging to the human species, as a common treasure-trove of all human beings, this problem would no longer pose itself. Sooner or later a planetary resources economy will become a necessity, and likewise a planetary sociological information-structure, whose perpetuation the former will make possible. Conversely, if matter and energy become the property of the planet and there is no longer any reason for them to be appropriated by a human grouping which does not know how to make use of them because it is not developed enough technically, then there is no longer any reason for failing to organise technical information itself on a planetary scale; there is no case for it remaining the property of a few human groups whose geo-climatic environment (the temperate zones) has favoured its development. But when people talk about disseminating technical information, they always think that the human groups who receive it must of necessity use it in the same way as those who fashioned it. Hence the imposition of a prefabricated culture on quite different ecological systems, however unsuited they may be to it. So far as we know, only China seems to have understood this mistake and to have used Western technological information in its own way, distrusting accelerated industrialisation in particular. Indeed, when technology is introduced it is not only the ecological framework which is disturbed by the harmful effects of an urbanisation ill-suited to the sociological structure; this sociological structure itself is turned upside down. The hierarchical grid of the industrialised countries is imposed on the former systems, the new hierarchical system being based on the degree of abstraction in occupational information. The finality of immediate survival in an often harmonious ecological equilibrium is replaced by that of production for profit, which is the basis of the industrialised hierarchical systems. And this production is futile, in the sense that the inhabitant of the country itself scarcely profit from it, all the profits going to the big international monopolies. All he gets is imported poverty. A few individuals are blessed with occupational training acquired abroad; they have been won over by a way of life which hierarchically gratifies them, and are conditioned to its prejudices and value judgements, which deprive them of the spirit of criticism. Lacking imagination, they try to apply lessons well-learned elsewhere in a geo-climatic and cultural framework not made for that end. And imperialism is only too happy to find interpreters to translate its ideology of cultural domination into the local language.

At all events, given free circulation of matter and energy throughout the world, and given free circulation of technical information, it would then become necessary to effect the opening of all the sociological information-structures. The liver, with its reserve glycogen, distributes glucose to the entire set of cells in the organism. What does it get in return? Energy substrates which, by means of a form of circulation which is adapted at every instant to the effort that must be made, allow the liver to conserve its information-structure and to secure its specific functions (see page 28). In this way it participates in the finality of the organic set, which is to maintain its structure as a set. Imperialism only secures the maintenance of its own imperialistic structure. And since its finality is

143

growth — production for the purpose of domination — the necessary result is the exploitation of the technically non-developed countries by the most technically developed ones. Now let us suppose that technical information itself can become the possession of the whole planet. This brings us to the problem we foresaw above: what is the finality of the largest planetary set at this moment? If it remains growth, then no solution seems possible. True progress will be made only if it is possible to equalise the advantages, and this certainly means stagnation or even regression for the technically developed countries, until those which are not yet technically developed have caught up with them. But the problem of hierarchical occupational domination will still not have been solved.

Moreover, another problem appears. With the free circulation of materials, energy and information, each social group will probably still want to set a value on its property in relation to that of its neighbour. This is what has happened so far, and it has happened in a completely unjust way which favours technical information and therefore countries with a high technical development. With the oil crisis, we have seen some of the exploited countries taking a belated revenge. Quite probably this revenge will continue, with increases in the price of all the materials which the technically developed countries have been taking from the technically non-developed. In due course the technically developed countries can exchange their technical information for raw materials and energy, instead of keeping it for themselves and processing it only for resale at a stiff price in the form of manufactured products. This will lead to the industrial development of the technically non-developed countries, who in their turn will embark upon industrial expansion. But what local information-structure will be in charge of this? Up to the present time only a few feudal interests have benefited, apart from the big international monopolies. The present energy crisis may accelerate the evolution of the sociological structure of these countries. However, within such a framework there is no likelihood of this structure being anything but a hierarchical one based upon occupational information, just as it is in the industrialised countries. But what will be the equilibrium in the exchange of raw materials, energy and information? The rules of the liberal economy, those of supply and demand, are likely to prevail. In this event, the technically developed countries will be handicapped by the high level of the needs they have obtained from their socio-cultural training, that is to say the habits they have acquired. On the other hand, their high technical development is going to push

144

them into accelerating their seizure of new sources of energy — solar energy and atomic energy, for example — and it will drive them to look for their raw materials (metals in particular) in the lumps that cover the ocean floor. And when they have got these, a new stage in domination will probably ensue: the countries which are at present technically non-developed, having reached the technical level at which we are now, will be left with their sources of energy and raw materials unused. The fate of countries possessing the classical sources of energy and raw materials may well be catastrophic. Even this dismal glimpse of things to come assumes that the technically developed nations of today do not use their armed might to obtain sources of energy and raw materials at any price, though this is unlikely, since the power blocs hesitate to clash, with the means of destruction as they are today.

Another factor to be taken into consideration, which shows the fragility of the concept of ecological property, is the fact that some goods are common to the entire set of human beings on the planet: the atmosphere we breathe and the seas around us. Now just as industries locally exploit and pollute assets which do not belong to them (the water in our rivers and the atmosphere of our towns and countryside), industrial civilisation on a world scale is changing the composition and temperature of the terrestrial atmosphere, charging it with radioactive elements, and it is polluting our seas internationally and destroying their animal life.

Finally, health technology has led to the disappearance of most of the great world epidemics. In the technically developed countries the increase in population which resulted from industrialisation has been moderated by the increase in the individual's economic well-being. But in those underdeveloped countries which were previously at an agrarian tribal stage, where each family was accustomed to losing many children because of the precariousness of child health, this protection has led to a population explosion and an ensuing rapid urbanisation which, in the absence of developed technological structures, has been the cause of catastrophic poverty. The absence of clear hierarchical levels has caused crisis to predominate over malaise, and generally speaking has led to the establishment of military dictatorships favourable to foreign imperialism, for the military profits from this situation as well as the dominant class. The countries of Latin America illustrate this process.

Consequently, for the planet as a whole as well as for the nation, the indispensability of the functional classes still seems to be the only indisputable power (see page 119). It may be objected that the

Eskimos and the Pygmies, for example, are fixed races and are not in a process of evolution; they do not seem to be indispensable to the evolution of the information-structure of a planetary organism, nor to its economic and energy equilibrium. They are shut away in their ecological museum, like the North American Indians in their reserves, and are destined to disappear. But if one is convinced of the importance of the genetic combinatorial, there is no proof whatsoever that they will not have a part to play in the future and that we would not then regret the disappearance of their accumulated informational and genetic pool. Every living form enshrines something that matters, not on account of sentimental memories of a past that is usually assumed to have been a happy one, but because of its potentialities for the genetic information of a future world. (It is true, however, that no one gets upset about the disappearance of Neanderthal man to the advantage of Cromagnon man.)

To summarise this particular approach to the problem of the world economy: here too, the possible objects of exchange are not only labour power which in any case is nothing without information) but also *mass* (raw materials), *energy*, and *occupational information*. These objects only have "use value" as a result of the socio-cultural training that teaches us what to do with them in the process of producing consumer goods. When the underdeveloped peoples have understood the importance of occupational information in the transformation of materials and the use of energy, there is no law saying that they must want to acquire it and evolve towards industrial civilisation like the rest. But if they are in fact attracted by technical progress, then perhaps a new stage will be entered from the point of view of the planetary information-structure. Indeed, the countries which are now underdeveloped will then be able to manufacture or, to be more precise, to "mechanofacture" their raw materials, to transform or use their energy and to supply the rest of the world with the product of their informed labour.

In this way they will acquire the indispensability of a functional class. (They do not yet have this indispensability, for while the industrialised countries currently have an endlessly increasing need for the wealth contained in the sociological niches of the underdeveloped countries, they can get on very well without the populations that inhabit them.) So far from accepting the disappearance of these peoples as a result of the growth of domination, just like the disappearance of the races mentioned above, we may note that their

146

opening can be achieved by their inclusion in the planetary information-structure. The opening in this case will be in a *vertical* direction. Their *horizontal* opening, that of association among themselves, is apparently difficult to achieve, but it seems to be the most effective and the most rapid way of achieving the desired vertical opening. Indeed, if they are separated from each other and included in sets with a high level of technical development, then even if they subsequently undergo a parallel technical evolution, this can only lead to the enforced adoption of a uniform finality, i.e. that of the industrialised nations, and to the abandonment of the special characteristics of the local sub-set. On the contrary, the first horizontal opening must unite the countries of the Third World in a new set which will have to discover its own finality. This should not be economic expansion in isolation; instead of reproducing the evolution of the technically developed cultures, they may be able to find another type of evolution which is original in character and capable of influencing ours. Perhaps evolution needs to wait for the complementary association of two kinds of structure in order to set off again towards a happy future, just as the symbiosis of primitive glycolytic forms with aerobic forms of more recent origin formed the first mitochondria and gave a new start to living systems in the biosphere (see page 32).

It is customary to note that species disappear because they are not adapted to new ecological conditions which have overtaken their niche; and people therefore tend to stress that the human species could also disappear, like the giant lizards of the Mesozoic period. People generally forget to mention that evolution did not stop on that account, and that what disappeared were lateral branches escaping from the main trunk, the phylum. I have just put forward the idea that every living form can be useful to evolution by virtue of the genetic potentialities it contains, and that we should be careful not to hasten their disappearance; but one says this in continuing ignorance of the mechanism and laws of the evolutionary process. Up to the present time the pressure of necessity has got rid of the awkward shoots, the systems which are closed from the information-structure standpoint. Let us, as the human species, fear subjection to such a pressure of necessity. The function of science is to try to penetrate the laws of necessity in such a way as to use them for the benefit of tomorrow's humanity.

I have already, in various contexts, spoken of levels of organisation. It is constantly necessary to ask which "organism" is being studied, which organised set is the object of our study, and what are the limits of its information-structure, at what level it is closed.

We have noted that a human organism, for example, is "self-managed". We said that the nervous system is not a "ruling class": it is only an intermediary, which is capable of registering consciousness of the variations occurring in the environment and, in response to them, of acting on this environment so that the organic set, whose welfare or distress it merely "expresses", may survive. It does not take decisions for the organic set, but simply expresses on its behalf the necessary decisions as to how it should behave if it is to find well-being or escape from distress. It acts in a system which, as far as the information-structure is concerned, is closed.

So if we seek the sociological analogy, it is certainly not at the level of the individual or of the enterprise that we ought to situate ourselves at the beginning, but at the level of the largest set which comprises all the human sub-sets. This largest of all sets evidently is the species and its environment, the planet.

It is this "largest set" which must manage itself if it is to ensure its own survival and pleasure. Any dissection within this set runs the risk of finishing up simply with the competitive struggles and pursuit of domination which I have described above. But let us examine whether it is possible, before we reach this kind of self-management, to pass over the heads of the national "blocs" and get at the national structure, to see if it can be self-regulating despite what we have just noted.

The nervous system is simply an intermediary between an organism's environment and its action on that environment in pursuit of the "desire" for the well-being of the whole body. This enables us to see that, according to this pattern, no centralisation of decision-making is acceptable. The only possible role for the central organisms in such a system is to inform the national set of the internal and external context of its activities, and to express the judgement of the national set in the action undertaken. It is the role of an intermediary: no more than that. If information is covered up to the advantage of the leaders, if there is any failure to disseminate information to the entire set of the nation, if there is any hindrance to the generalisation of culture by which every individual and every functional class can express their views, and in particular if information is directed from the top down, with decisions being pressed upon the base, then the result will be not self-management of the national set but either pseudo-democracy or a bureaucratic system. No individual and no group of individuals has the authority to make decisions about the well-being of the set, and if they have to invoke the ignorance of the masses when they decide things on their behalf, it is because they have not fulfilled the task of disseminating "generalised information", and are restricting themselves to the dissemination of the specialised occupational information required for growth, profit and the maintenance of domination. It is therefore necessary to invent a national information-structure and hence also a form of social organisation which is capable of helping the flow of circulating information.

If we explore the analogy between the individual organism and the social, national organism more deeply, we note that the environment of the latter is first of all the territory which constitutes its immediate ecological system, its environmental niche, but that socio-cultural and economic relationships are also established outside this territory with other national organisms, with whom materials, energy and information are exchanged. What autism is to the individual, autarchy is to the nation. But in the case of nations, as of individuals, we may observe the pursuit of domination and the establishment of value hierarchies, based upon the specialised information from which productivity springs. This domination, which is a consequence of the closure of the national system, is of course secured essentially for the benefit of the dominant class, and it is this class that stimulates commercial and industrial competition in pursuit of new outlets and markets. But the national set often takes the cause of the ruling class to be the cause of the nation, and regards the defence of the interests of the dominant class as the defence of all the citizens. It is interesting to note that we seem here to be accepting the idea of classes in the classical marxist interpretation, which we previously criticised. The classical interpretation is backed up by the fact that this ploy has not yet succeeded in bringing the antagonistic class, the working class, into the fold, since the capitalist class seems to have effected its opening horizontally. Indeed, capitalism is a transnational phenomenon; capital has lost its nationality, and those who possess or manage it behave in a way which has nothing to do with the national interest but everything to do with the accumulation of the said capital. But in reality this does not mean that there exists a precisely demarcated capitalist class, a set of individuals presenting similar

149

functional characteristics. We have already had occasion to remark that the bourgeoisie and the proletariat are distinguished much more by subjective experience than by socio-economic criteria. This is due to the fact that a person "feels" bourgeois or proletarian according as he "feels" sufficiently gratified or not by the hierarchical system which is founded upon occupational information. Every nation today, whether capitalist or socialist, possesses its own hierarchical system, and only the labels are different. The distinction is due to the fact that in one case capital has its own existence and constitutes a finality pursued by the whole hierarchical system, whereas in the second case capital is a secondary notion (since in principle it is collective) but the hierarchical system has its finality in itself. Perhaps that is why state collectivism has not yielded the expected results as regards economic expansion, for the motive of domination does not lie in the economic field. But in reality, if the situation is closely studied, there is little difference: the direction in which surplus value is used, in one case as in the other, escapes the control of those who produce it, and hierarchical alienation is still to be found. There is no dominant national class which could be more or less integrated with and horizontally open to the dominant national classes in the capitalist countries. But international capital certainly exists, just as the human behaviour that makes this the finality of its actions exists. Every human being who accepts the ensuing hierarchical system, which is based on the degree of abstraction in occupational information (since the accumulation of capital cannot proceed without economic expansion and hence the development of occupational information), is a bourgeois. To pursue the social advancement institutionalised by this system, and hence to accept it at whatever level in the hierarchical scale a person happens to be, is to be a capitalist even without owning capital. In the same way, to try to rise in the ranks of the party, in the hierarchically privileged organisation, is to turn power into capital or the "surplus value of power" (as defined by Gérard Mendel[42]) yielded up by those lower down the hierarchical scale.

It can be proved that pursuit of domination is the real motivation. For while the boss has gradually accepted improvements in the workers' conditions (which in any case he needs, for the growth of purchasing power and hence for the growth of profit), he has generally refused and still refuses to share his decision-making power either with his staff or with the workers. So it seems that the problem of growth has from the outset been firmly based upon the pursuit of domination, even if this is less and less achieved by means of the ownership of capital. If we take active and effective trade unions into the reckoning, the capitalist economy which guarantees "growth and expansion" may be better than a bureaucratic régime at giving the labouring classes some sort of material welfare by means of the random and obligatory increase in consumption. But in fact such a system, even with the countervailing influence of powerful and effective trade unions, does not in the least improve the labouring classes' decision-making power (or rather, as we have already suggested above, what is *thought* to be decision-making power).

So the point is not for the dominated classes to replace the dominant classes in making decisions that are always about producing more and more commodities to increase profit and investments, which actually is very adequately done without their taking a hand in it. Certainly the point is not to improve working conditions, free time and incomes at the expense of these investments and of growth, unless the stage has been reached where these demands can put a stop to growth by suppressing both the basic motivation of profit for domination and the investments which increase production and hence profit. The ultimate decision will concern a choice between the production and possession of more and more consumer goods, and the creation of a way of life that allows everybody sufficient free hours every day to increase their knowledge of the relationships between human beings in all their forms: biological, psychological, sociological, economic and political. Clearly, this way of life would considerably reduce production, or at least would stop growth and expansion. It would result in the condition which is dreaded so much by economists and politicians: stagnation. But with stagnation in the production of consumer goods a real "miracle" could blossom — not an economic miracle but a human one. The question would be sharply posed as to whether the finality of the human species on this planet is always to make more and more commodities, or to understand more about both the inanimate world and the world of living creatures, including the human world. For the first time this problem would no longer be left to a handful of philosophers but posed as a question for the entire set of human beings, with enough information diverted away from winning power to enable an answer to be found. The first result clearly would be the sudden disappearance of value hierarchies. If we come back to the level of the national sub-sets of human beings, we encounter once more the idea that the information-structure, sociologically speaking, depends on the finality of the set. Is the individual in this sub-set only a machine for production, or can his action on the environment (the finality of which is his "well-being", his pleasure) also be expressed in some other language than that of economics? In so far as machines have neither entirely replaced man's economic functions nor become self-reproducing and self-improving (a state of affairs that won't be reached in the near future), there will continue to be productive activities for human beings to undertake, even if within this productive activity thermodynamics increasingly gives way to information. But it is implicit in this that if the finality is not the growth of production, free time will continually increase in relation to a fixed amount of production. Later on we shall look at how this free time might be used, for a problem will arise in our present day civilisations similar to the problem which arose at the beginning of the neolithic age. When it became possible to build up reserves and to cease being subject to a daily hunt for food in order to survive, a considerable amount of free time was then available which neolithic man used to develop his technical specialisation and to study the laws of the inanimate world in greater detail — he did not merely give himself up to leisure and games. Physics, thermodynamics, mathematics and craftsmanship had their origins in this period. The shaping of matter in order to provide effective aids for survival also began at this time. The use of energy was still empirical, and was to remain so up to the industrial revolution. This last stage is also a quite fundamental one, for it permits the transition from knowledge of the physical world to knowledge of the living world. Thus the free time which is recoverable as soon as humanity's finality is no longer economic expansion alone (the consequence of using information solely in the domain of matter and energy) can be used to obtain knowledge of the living world as a whole and of what animates the behaviour of each of us in the social situation.

What then is the information-structure of tomorrow's national societies? The hierarchical structure of our epoch will have less reason for surviving, since the expansionist finality will be less powerful. Even if this hierarchical structure is retained in the productive aspect of human societies, it must at least begin withering away once the importance of the productive aspect itself becomes less exclusive. What will still need to be invented is a system of inter-occupational relationships acting as chains of servo-mechanisms, and the means for a gradual inclusion of systems whose finality will not exclusively be production for production's sake but will involve the "human sciences" (though not merely in their discourse-related forms).

But even in the aspect which is still productive, one may already imagine a system of relations which is no longer hierarchical. Every human being in isolation is forced to belong to a hierarchy, and within it he adopts paternalistic, racist or dominant behaviour towards his "inferiors" and childish or submissive behaviour towards his "superiors". I have drawn attention to the solidarity of the system. In order to open the closed system of the individual on to a global set which is not hierarchically organised or "individualised" (for one must divide in order to rule, and the more the hierarchical system is staggered and individualised, the greater the sovereignty of the commodity), it is also necessary to open it on to a "functional set", a functional class.

But for this it is necessary to submit to the existence of levels of organisation, to each social class: "class" in its broader meaning, rather than that of the opposition between capitalist and proletariat. This broader meaning is based upon the analogy of *function*. In a single organisation or enterprise where everybody works together, a certain number of classes will exist whose individuals will feel in solidarity with each other because they fulfil the same function. This is what constitutes the "functional classes". But for this a certain number of transformations are necessary, and it is also essential to do away with private ownership of the means of production.

The first, the most fundamental and certainly the most difficult transformation to achieve is the one which I have already stressed again and again: the disappearance of both paternalism and infantilism in inter-class relationships. At all levels of organisation and for each individual, it is difficult not only to acquire class consciousness but also to shake off psycho-familial paternalism towards people from the "inferior" class and infantilism towards those from a "superior" one. It is hierarchies of value that poison social relations.

Hierarchies of value at present are established, as we have seen, on the basis of specialised technical information; self-management must necessarily lead to their disappearance, if it is to enable the power of the functional classes to emerge. It is therefore probable that hierarchical technological domination will oppose the disappearance of its own power. So there is a tendency for the class struggle to be not only between "proletarians" and "capitalists", the owners of the means of production, but more importantly be-

153

tween the proletariat which is deprived of technical knowledge and the technocratic bourgeoisie. In contrast to previous experience, a technocratic and bureaucratic class may take the place of the traditional bourgeoisie as the bourgeoisie took the place of the aristocracy. The gentlemen of the technical and management staff who, as everyone knows, have their "responsibilities", will oppose any challenge to their authority from ignoramuses. But they do not challenge technical knowledge when it is sound; what the latter are challenging is the way of utilising these technical achievements, and the authority (which is technical) conferred by these achievements in a context of hierarchies of value, wages and power because the technical achievements fit in with the pursuit of domination by expansion.

So let us return to the problem of the national level. In a world which continues to be guided by productivity for domination, a self-managed nation — by virtue of the persistence of the global finality — will probably achieve only a semblance of self-management, which will in reality be technical and bureaucratic management. The only way to avoid shipwreck on these rocks would be to supply the generalised information which would permit the establishment of an opposing structure consisting of functional levels of organisation; this would give power to the functional classes and set them against the technically endowed value hierarchy as it exists at present. In fact I have already stressed the fact that technical supremacy disappears when confronted by generalised information. So it seems absolutely clear that in order to get rid of the hierarchical domination which is based on technical knowledge, it is indispensable to see that non-technical knowledge is generalised, thus enabling the power of the functional classes to be installed. (Would this not be what is referred to as the generalisation of culture? It would not of course be the culture of a class; the latter has usually been a tasteful but useless decoration, a means of distinguishing oneself within the technically developed hierarchy of value.)

The *"endonational" organisation** of social structures, therefore, requires generalised information about its external as well as its internal environment. Without such a system for conveying information the national organisation is incapable of acting in a

---

*"Endonational" is a neologism, used to describe an organisation which is "internal" (endo) to the nation. This implies that "inter"-national organisations are simultaneously both "endo"- and "exo"-national in relation to a given nation [H.L.].

coherent way on the environment, nor can it define its internal equilibrium. It can only lead to class power, whose first thought is to satisfy itself and to maintain its hierarchical domination. Generalised information is about the productive activity of the social set or the aspect of its activity that is in some way related to energy, i.e. its metabolism: but it also has to do with the very structure of the organism and the relationships between its elements, from which its overall productive activity derives. Here we encounter once again the idea of levels of organisation, which will be familiar to the biologist. Starting with the individual, whom we have already seen as an analogy of the cell, groupings of these individuals made up of organs, tissues and systems find their own analogy in enterprises, industries and large-scale national activities, whose set unites in the effectiveness of its global action. Each level of organisation regulates and controls the level below it, but each level, as in an organism, is indispensable to the activity of the whole set. Its finality is indeed its own personal satisfaction, but this is only realisable through the satisfaction of the set, which in turn is possible only because of the effectiveness of each level of organisation. Of course we are dealing with regulated systems, but systems whose information comes from outside through the establishment of servomechanisms (the choice of the global finality is a result of the pursuit of satisfaction by all the elements). This is how an organism functions. Here again we encounter the idea of generalised power, but now it is even more evident that this generalised power is possible only if there is also a generalised information which permits action at every level of organisation. Now we have seen that generalised information is only conceivable if it is not subject to direction, if there are no specialised hierarchies of value, and most particularly if there is time to receive the information and to seek out its sources. Generalised information will require a permanent distrust of logical analysis, which is simply the cover for an unconscious emotional predisposition to domination; it will also require the abandonment or at least the critical questioning of simplistic and reassuring grids, pseudo-scientific certainties and constant references back to the great figures of the past.

Within this kind of national organisation, the power accruing to a functional level of organisation within a sub-set — an enterprise, for example — will clearly depend on the necessities and objectives of the larger organic complexes. We can describe this interdependence between levels of organisation of growing complexity as a *vertical* system. In such a system, any external command at a given

155

level of organisation is not a hierarchical command but an informational one. This is a result of the pursuit of a global homeostasis, which is forced to take into account the "desires", that is to say the structural and energy requirements of each element in the levels of organisation lower down, as well as those of the organism as a whole whose global action makes it possible to meet these requirements. But we may also imagine power coming from a *horizontal* structure. By this I mean a power which springs from the indispensability of elements distributed among different organs throughout the whole of the organism but united by a similarity of function.

The set of teachers in a nation represents a social class, as does the set of those who are taught (to refer back to Gérard Mendel's example). But these two classes also exist in one and the same institution (in the same organ), at the level of the system (national education). By defending its power in the vertical and horizontal direction, each social class will be able to take part in the global homeostasis and curb the specialised hierarchies of value.

The horizontal and vertical structures have the advantage of enabling us to see the possibility of an "opening" of the system. Each class, each level of organisation is a closed system. Its transformation would only be possible by means of a transformation of its "function" (since here we are dealing with functional class power). As each student grows older, he will leave his class and be replaced by others, but the class of students will remain as long as one generation transmits its experience to the next. In saying this, no assumptions are made about the way in which students are taught (whether it is paternalistic or otherwise) nor about the way in which the transmission of experience is realised; nor are any assumptions made about the recognition (or lack of it) accorded by other classes to the class of students, whose indispensability is as evident as that of the patient in the hierarchy of a hospital. But as in every regulated system, this level of organisation is closed upon itself, that is to say the effector (the class of students) possesses an effect (acquisition of experience) which acts retroactively on its factors (teachers and other functional classes). Its opening results from its membership of a system in which other levels of organisation take part. We have already dealt with the *limit* of national integration. The opening at a higher level, the *vertical opening*, consists in the integration of the national organism and of the level of organisation that it represents into an international set. The *horizontal opening* is that which results from the different functional classes meeting their

156

analogous classes from other national organisms. We are beginning in a tentative way to see this kind of meeting in the Common Market. But here again, these functional meetings are only possible in the absence of any fundamental antagonisms between the national organisms, that is to say when the group of nations tending towards integration has a common finality. The principal obstacle to vertical and horizontal opening is therefore, as always, the existence of hierarchies of value and of domination.

If we consider this at the level of the enterprise, we can compare the latter to a thermostatically controlled water-bath in a laboratory (see page 22 above.) We have an "effector" whose "effect" consists in a certain kind of production, thanks to certain "factors" (raw materials, machines, sources of energy). It is a regulated system, if the scale of production fixed by the system itself determines the value of these different factors. Thus it is a closed system. It becomes an open system if the external instructions reaching it (which are the result of this enterprise belonging to a larger set — the industry — which itself belongs to a still larger set — national production) supply it with the information which enables this regulated system to function with a certain intensity, at a certain level of production for example. Then it is an open system in the sense which I have termed *vertical*. But alongside this vertical opening, i.e. its membership of a larger set, the enterprise can also possess a *horizontal* opening by virtue of the fact that it joins up with other enterprises whose activities are either analogous or *complementary*. The largest set is then a consequence of the *meeting* or intersection of different enterprises, and may be seen either from the *thermodynamic* standpoint of productivity or from the *sociological* standpoint of the functional classes which give these enterprises their social structure.

In such a system competition between enterprises no longer makes sense, since their production answers to information coming from outside — and this certainly represents what we would conventionally call planning. This must not be imposed from the top down but accepted at the cellular (individual) level, because the information necessary for the effective functioning of all the levels of organisation, including the national and international set, circulates at this level.

Let us take a specific example. A cork factory comprises a certain number of individuals who each perform a specific function in the enterprise, but who meet together in functional groups (quite apart from any occupational hierarchy) if the circulating informa-

157

tion is circulating properly. (We shall see later what this circulating information consists of.) This *sociological* aspect represents the *information-structure* of the cork firm. It also has machines which fulfil a particular role. These machines are an extension of the human being, sometimes even replacing him, though without thereby possessing hierarchical power (which, incidentally, shows that a function can be fulfilled without its being supplied with hierarchical power). They are likewise part of the information-structure of the cork factory. Like a human being, they need energy in order to perform their function. Man finds energy in the form of food and, more broadly speaking, in the form of everything necessary to satisfy his basic needs. For the machine, this energy can be hydraulic, electric or from fossil fuel (oil, coal, etc.). The role of these machines, like that of the human being whom they replace or "improve", is to transform crude raw materials into a product that is more elaborated and more informed. This information is supplied by the technicians who conceived their structure and programmed them; these technicians rarely ever belong to the personnel of the factory but are members of another functional class, in a relationship of complementarity. The energy and the raw materials necessary for the effective utilisation of machines and men represent the *economic* aspect of this enterprise. The cork factory as *effector*, whose structure is constituted by machines and men according to certain relations which determine the particular nature of this structure, produces a certain *effect*: corks. For this, certain factors are indispensable which are always the same: energy, mass (raw materials) and circulating information. Of course this effector can function as *regulator*, but in this case the circulating information coming from outside the system would be useless: either the quantity of information and of raw materials such as cork-oak would limit the production of corks (feed-before) or, conversely, if the production of corks was fixed once and for all at a given value, this would regulate the quantity of energy and raw materials to be used (feed-back). The finality of this factory, which is making corks, the finality of this closed structure goes on to become included in another finality, in another structure which then enables it to become open. In reality, however, this is not always the case. If the production of corks is only a secondary goal, the primary goal being to make profit and to increase the scale of the undertaking so as to make it a monopoly, one can imagine that good advertising could make people consume corks not only to cork up bottles but, for example, to make necklaces, earrings and

158

fashionable but utterly useless items such as the kind of mobiles which let the user cut them to his own taste and develop his personality, etc. etc. In this case we have a closed structure, a veritable tumour growing out of control which, however, will always find a logical defence of its existence and productivity (since it is a human enterprise we are dealing with). Thanks to the development of the cork industry a greater and greater number of workers are going to find work, not only in the firm, but in the factories that manufacture cork-making machines; the GNP is going to increase and, with good publicity overseas, growing exports of corks will bring in foreign exchange, and the flaming torch of our national genius will be borne aloft. A sizeable market can be developed in the underdeveloped countries, where cork necklaces could easily take the place of those made of shells. If the boss of this cork firm builds some canteens for his workers, or some day nurseries and recreation grounds, and if he pays them properly, he will be a good boss. His social success will win him a knighthood or at least an OBE, especially if he knows how to get a few local politicians on his side. But we have seen that instead of such a closed structure, we could imagine an open structure. In this case the regulator, the structure closed upon itself, will be transformed into a servo-mechanism. The corks will be used — for example — to seal bottles. The bottle industry will inform it of its needs. The bottle industry will not give it any orders, but will simply let it know that the bottle industry is expanding or contracting, and has need of a given quantity of corks per month. But the bottle industry can be a closed structure in just the same way as the cork industry, and in this case it would be making bottles in order to sell them against all comers. But it too can become open, by being included in larger industrial sets which produce liquids to put in the bottles, etc. In both cases the closed structure opens on to a more complex set, by means of a servomechanism of information from outside the regulated system. The supply of energy and raw materials is no longer governed by motives of profit and the expansion of the system, but by a function which is part of the function of a more complex set. This is only possible if the finality of the larger set, which we have considered in this case to be a national finality but which can be extended as far as the whole species, is not itself based upon expansion, the increase of profit or the protection of occupational hierarchies of domination. It may be objected that supply and demand, the desires of consumers for corks, bottles or liquids etc. are the basic precondition of this expansion. Unfortunately, we know that man expresses a desire only for what he knows — paleolithic man had no desire to own corks. By the same token, the hierarchical structure with its occupational base (and the information-structure at all levels of organisation) will depend upon the finality envisaged by the system as a whole, and upon that of all the sub-systems which participate in its functions.

In a hierarchical system of the kind I have described, circulating information has no need to circulate. The finality being to make the greatest possible number of corks in order to make the greatest possible profit, the boss and the board of directors think they are taking decisions. In reality, all they decide is what allows the production of corks to grow (they study the market, the possibilities opened up by advertising, the supply of raw materials and energy, workers' wages, profitable investment in machines, etc.). The kind of circulating information which actually needs to be disseminated to the entire set of the human information-structure of the firm, concerns all the problems of a general kind which I posed at the outset of the argument, especially ideas such as structure, information, sets, open or closed systems, the finality, sub-systems and systems of which the bosses themselves are ignorant since they are working in a closed system. It also concerns the relationship between structural ideas and those of mass and energy, and those connected with the biological mechanisms of human behaviour in the social situation. This contribution is far more fundamental than any purely occupational experience of a more or less specialised character, for it is the basis of a kind of political behaviour: the opening of the closed system constituted by the individual organism in a functional group which is itself open vertically and horizontally in social sets of increasing complexity.

Clearly, the economic liberalism generally defended by those whom we conventionally call "conservatives", that is to say individuals who declare themselves fundamentally opposed to anarchy, represents the very prototype of the anarchic system. I am well aware that this system claims to be governed by rigorous laws, with production being governed by consumption and supply being governed by demand, and a harmonious equilibrium thus being established. But experience shows that this equilibrium does not exist. Demand depends upon publicity and therefore upon supply, and the desire for objects is governed by commercial information and invested capital. Economic anarchy establishes itself, and the strongest impose their domination. What is called democratic planning is in fact swept away by the powerful blast of profit.

The individual is unaware of the determinisms which steer this mad society (mad because it is not geared to reality); he is blinded and deafened by the automatisms and needs which it creates in him. He is diverted from the fundamental problems by sub-problems whose importance is blown up either as a result of the very ignorance of those who direct things, or of conceptual solutions which are not only out of date but also themselves manipulated by other hierarchical systems of domination. The individual allows himself to be carried away unresisting, in comfort or in discomfort, by this same powerful blast of profit. If individual profit is no longer linked to that of the enterprise, if it becomes the collective profit of the state, it might appear that since the motivation has changed, the economic anarchy and absurdity of the previous system must disappear. But experience shows that in reality things are otherwise. This is not because there is no longer any motivation at all (although this certainly is the case for many individuals in such a system). The motivation which dictates the pursuit of domination still exists. All the links between the individual or private good and the collective good exist, for the collectivity assumes various dimensions: local, regional and national. What seems to define the collective good in a collectivist system is the fact that it is obtained by means of collective capital. However, its mode of utilisation does not usually spring from a collective decision but from that of the bureaucrats or parliamentarians who allocate state funds. Consequently there is such a gap between the individual's labour and the collective gratification that labour ends up by yielding surplus value to the state and is unable to exercise direct control over how it is used and distributed; its motivation must therefore be considerably reduced. And the income handed back to each individual by the state is allocated according to a hierarchical scale. This, as everywhere in the technically developed societies, is based on the degree of abstraction of occupational information. But it is also based on submission to the rules of the system, on a knowledge and strict application of the ideology, since this is what permits domination and the maintenance of the hierarchies. Even if the scale of wages and salaries is much less differentiated than elsewhere, still it exists. But since it is much harder to use them for consumer satisfaction, the scale of wages and salaries has less need to be staggered. Only a small number of privileged people in the régime benefit in this particular respect. It is easy to understand how in such a system, despite a reduction in the production of useless goods, the absence of motivation for individual acts of grati-

161

fication may be a hindrance to overall production.

The finality of the system remains the same: productivity. Since this productivity, directed in particular towards collective assets and national defence, is felt to be insufficient gratification for the individual, and since the individual finds that his consumption is as uniform as his socio-political thought, the profitabiltiy of the system does not appear to be improved by the disappearance of the individualistic profit motive. Its place is taken by the search for promotion in the hierarchy of the bureaucrats who dispose of power. But what kind of power?

Clearly, a generalised political power should be based upon a generalised political knowledge. The political power which might eventually belong to the functional classes must steer clear of two pitfalls: corporatism and the cult of personality, at the level of even the simplest groups.

"Corporatism" consists in trying to impose the political power of a functional class or more often an occupational class without taking account of the overall structures in whose constitution this class participates. Corporatism cuts the class off from the servomechanism of the informational source coming from outside the closed system constituted by the class. The corporation is the set of the results of the closure of a system. It is a veritable cancer which thinks only of its own pleasure, its own well-being and its own satisfaction; it relies upon its own indispensability, without recognising the indispensability of the other functional classes. Within one and the same functional or occupational class, hierarchies of power exist which can find openings in a horizontal direction by associating themselves with similar hierarchies of power in other functional classes. By actually seeking such horizontal openings, the less favoured hierarchical levels within an occupation try to curb the intra-occupational hierarchy in a corporation. In order to curb corporatism, value judgements and the isolation or sporulation of occupational sub-sets, it is necessary to look both for the *vertical* opening in the larger sets encompassing them, and for the *horizontal* opening of each level of organisation, of each sub-set of the structure.

The "cult of personality" is not far removed from the paternalism and infantilism to which we have already referred. It involves the surrender of political power to one individual or a handful of individuals able to impose their personal views on other groups, who accept this imposition because it gives them a sense of security to believe that certain people are better informed than others about

162

the socio-economic complexities and are therefore better equipped to act. The generalised information of the group is imperfect, either because there is no time to acquire it, or because the dominant personalities deliberately hoard information and only let the group have what allows them to maintain their domination. Here again we encounter the problem of generalised information. We also encounter the fundamental idea that political power can only be the expression of a functional human group, and not of an individual, and that the political action of a national set can only be the expression of the political power of all the functional classes participating in the strucure of the nation. Clearly, this comprehensive political power must be established by appealing to the vertical and horizontal opening of human groups and national subsets. One cannot imagine a single cell imposing its power on the set of cells in an organ or an organism. When one of them goes it alone and begins multiplying on its own account, it gives birth to a cancer. But usually when a cell suffers such a fate in a normal organism, it is rapidly throttled by what we conventionally call the "means of defence", which destroy elements that do not conform with the common destiny. It should be emphasised that this is not submission to "conformism", for a normal organism is capable of evolution. But evolution takes place at the level of *informational enrichment* and not of regression to a unitary egoism. Memory and the enrichment of experience are characteristics of living systems. But this acquired experience is used to satisfy the organic set and to improve its conditions of life. The heriditary memory (that is to say, the genetic structure), the immunological memory and the nervous memory are no different. If certain molecular or cellular groups fulfil these responsibilities, they do not do so on their own behalf but on behalf of the organic set.

Here again, the distinction between information and thermodynamics is crucial. When I speak of "improving the conditions of life of an organism", I refer essentially to the maintenance of its biological equilibrium in relation to the environment, by means of its action upon that environment. Thus an organism does not improve its conditions of life when, given a certain quantity of energy which it releases and which is required from it by the resulting variations in the environment, it eats more than the quantity necessary to re-establish the constants in its internal environment (for example, the concentration of glucose in its plasma). If this happens there is a disturbance in the metabolism, that is to say in the functioning of the microscopic chemical factories — which is what cells are. Illnesses appear: obesity, diabetes, arteriosclerosis etc., diseases of excess, consumer diseases which accelerate destruction and death.

A consumer plethora exists in human societies too, and I have tried above to show the mechanisms behind it, which are based on the desire for power when all that actually needs to be secured is the satisfaction of basic needs. Assuming that analogies may be drawn between social and human organisms, growth and expansion must have limits. Unfortunately, while we have quantifiable standards for controlling the biological equilibrium of a human organism, we do not yet know what thermodynamic (nutritional) constants apply to social groups. How much energy should they take from the environment, what kind of social equilibrium (what kind of structure) should this maintain, what labour should it supply, what quantity of mechanical energy should it release?

Here we touch on the roots of the analogy. When we talk of maintaining biological equilibrium, which is a factor in what we describe by those dangerous terms "pleasure" and "well-being", it is in fact a question of the *maintenance of a structure*: the maintenance of the relationships and precise connections existing between its constituent elements, at all levels of organisation in an organism from the molecules to the organism as a whole.

If the *social structure* rests on the existence of relations between dominant and dominated and on social advancement, which in turn rests on hierarchical promotion in an occupational framework that creates consumer goods rather than new structures based upon knowledge of what we are, then the means of reaching this aim may well be expansion and apparently endless economic growth. This is how the diseases of excess arise, as well as the socio-economic and ecological disasters which we are beginning to observe.

At the end of a long period of biological evolution by trial and error, the human organism is still incapable of profoundly influencing its own structure, which has so far been effective merely in securing its survival in the biosphere. Is the social organism as we see it today just a biological slip-up, like the giant lizards of the Mesozoic period, and is evolution only possible if the imperfect structure founded upon domination disappears? Or rather, since structure is indissolubly linked with the finality of its actions, is it not this finality that ought to be changed?

For the moment our analysis will continue to be based on a national social structure (though this might just as well be conceived as a regional structure) and on an information-structure which uses mass and energy. On this basis, let us consider the immediate ecological framework on the one hand and its informational and economic relations on the other.

For a long time the immediate ecological framework has enabled human societies to survive. Although the exchange of raw materials and of manufactured goods was possible from the beginnings of the neolithic age (and sometimes over long distances), many human groups, particularly in the middle ages, could split up and live in autarchy because the scale of their needs was very low — it was limited to the basic needs for immediate survival and the fear of invaders and nomadic warring hoards. The spectrum of needs developed and became broader as a result of the socio-cultural diffusion between groups, and the ecological niches of individual groups became insufficient to meet these needs. The exchanges increased. With the growing importance of capital in the establishment of domination, and then with the lightning development of technical information in the industrial era, the need for raw materials and energy became dominant. The ecological niche quickly became insufficient to meet these needs. Technical information then enabled the exchange of manufactured goods and their production in large quantities to increase: this was the principal strength of the industrialised countries. The latter entered the era of colonialism, which allowed them to slake their thirst for raw materials and (with the discovery of oil) for energy.

Thus the national information-structure can be considered in relation to its immediate ecological niche and its structure within this niche, in so far as it finds raw materials and energy there. This is the situation of many primitive human groupings which are in ecological equilibrium with their environment. Their lack of an opening on to a wider set determines their evolution and generally results in gradual destruction and death. For all the sympathy we can feel for the simple and "natural" conception of human life, we must challenge the implied value judgement and remind ourselves that a social structure can only evolve by being included in a more complex structure. But of course, everything will depend upon the type of exchanges which then take place between the structure of this sub-set and the set to which it is going to belong. There are exchanges of information which transform its structure without taking away its originality, i.e. the specificity of its function. And there are exchanges of materials and energy in a more or less crude form, more or less transformed by the human group under

165

consideration: i.e. materials and energy on which the information supplied by the human group will already have played a part, turning them into products of human industry.

The only thing that a human group can exchange with others is, in the last analysis, information. Raw materials and energy can be "stolen" from a group which does not know how to transform them. Finally, for a theft to take place there must be property. There is no basis in biological principles for the ownership of any element in an ecological niche which the human group living there does not use to ensure its own survival. The evolution of the non-industrialised world will perhaps not consist in increasing the exchange of raw materials and manufactured products on the one hand and cash on the other, on a better balanced basis. It rather suggests the transformation of its natural resources into highly sophisticated products, by means of an exchange of technical information. This is what some of the Arab countries seem to have understood recently, with regard to the oil which their ecological niche contains. At all events the opening of closed systems at any level of organisation is, in this perspective, directed only towards the production of consumer goods, which continue to constitute the finality of the largest set, the species: energy and raw materials are processed by information only in order to serve consumption. But what if information became an end in itself: the information-structure of human societies, and the circulating information? What if materials and energy became merely the means for increasing the knowledge and the evolution of the structures? What if human society became an informational society?

As a conclusion to this chapter we might perhaps look at the problem of inflation, within the framework of ideas that we have been developing. A human society, established in a particular ecological system and drawing from it the essentials with which to satisfy its needs, can only avoid inflation if the labour and information which it takes from its ecological niche in order to transform its raw materials and energy (without going elsewhere for them) enable it to maintain its information-structure in a "steady state". But as long as the information-structure consists of a hierarchy of domination obtained by the accumulation of capital, and as long as this hierarchy has recourse to commodity production and the increasing abstraction of exclusively occupational information, it is absolutely essential that these commodities should come on-stream. To sell them, somebody must buy them; the greatest possible number

166

of individuals must take part in the purchase. Hence the importance of advertising in this type of society to make the goods known and animate the desire for consumption. To buy, one must have the necessary money; consequently wages must rise. But when wages rise, the margin of receipts over costs — the capitalist's profit — diminishes, so prices rise. Let us now suppose that the social group under consideration is no longer satisfied with what it finds in its own ecological niche, and goes abroad to search for the raw materials and energy it needs for the growth of its own production. The situation becomes complicated, since a counterpart for what is taken must be supplied to those who occupy the other ecological niches. We have seen that the problem is simple for planetary imperialism where non-industrialised peoples are concerned. But between national groups which are highly industrialised, exports become necessary. Obviously this simply defers the problem, favouring first the most technically developed and technically best informed groups, and finally generating the endless race between prices and wages on an international scale, with wages always lagging behind prices so as to enable capitalist profit to grow. So in order to struggle at the national level against inflation it is necessary to export, at the point where the margin between receipts and costs necessary for the purposes of investment and expansion becomes static. In the latter event, it is not the national group that secures the growth of sales to maintain profits, but the other national groups; national inflation can only be kept down by selling more than one buys and therefore exporting inflation. In effect, if the importing countries are not to see the profit margins in their industries diminish, they must sell more and at higher prices within the framework of their national economy. Consequently they will have to raise wages, and so on. This is probably only one very general aspect of a problem that immediately becomes more complicated by virtue of the existence of international banking funds and all the speculation that comes from the existence of these funds. But it shows that inflation in the "capitalist" countries can only be controlled by the more or less efficient but unfettered exploitation of the non-industrialised peoples, for inflation seems to be synonymous with expansion so long as one remains within a closed system.

In the socialist countries today, the problem has little chance of arising. In fact demand does not govern production, since in the absence of advertising neither the desire nor the cultural need for consumption are created. On the contrary, supply is at the source of

167

demand, and supply is the vocation of those in power, since production is entirely planned. Production may increase as a result of increased mechanisation. But it is far more difficult to diversify it, since there is no incentive for competition between types of products, only between amounts produced. This system could have been effective if not only the incentive of profit but also the incentive of domination could have been done away with, and if the masses could have been given new incentives. Now these have been restricted to a fixed ideology, and to the exploitation of this ideology beyond the national frontiers so as to establish national domination. Which human group will prove capable of adding to the achievements of marxism something that could quite fundamentally transform them, namely the concepts derived from biology in general and the biology of behaviour?

Finally, let us try to sketch the contents of this chapter diagrammatically. Although occupational activity is not the only activity by which an individual is characterised, let us first sketch a human group engaged in production: an enterprise (see figure 10). We shall obviously treat it in this case as an "effector". This effector possesses a structure, an "information-structure". It consists of human beings and materials. The human beings are organised in hierarchical relationships, with which I have already dealt at length. There is also a wild proliferation of hierarchies outside the framework of the enterprise. Note in passing that this organism, the "enterprise", does not change its hierarchical structure despite being exposed to a constant "turnover". It is like a living organism. In a living organism, the "turnover" is characterised by the unceasing replacement of atoms and molecules without the structure of the set being thereby affected. Similarly in the enterprise, the constant replacement of the individuals who make up the labour force will not generally speaking affect the structure of the set. This labour force is assigned to operating the "material" — machines which themselves wear out and will have to be replaced.

This set, which is a closed set as far as its structure is concerned, is open in terms of mass and energy. The individuals who make up the workforce need food to maintain their living structure, as well as clothes and shelter for themselves and their family. The energy necessary to keep the individuals of an enterprise in good working condition is supplied by the mediation of wages. The maintenance, replacement or enlargement of the machines is looked after by investments. This "effector" enterprise maintains its struc-

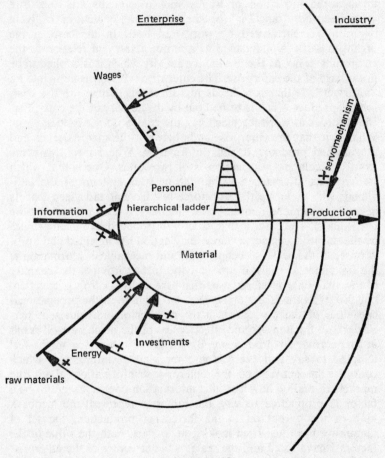

Figure 10

169

ture (this is the sole and fundamental finality of every living organism) by means of an action upon the environment, *production*, which represents its "effect". But for this it needs "factors", which again consist of mass and energy. As in the case of a living organism, mass and energy guarantee the stability of the structure through the mediation of wages and investments and therefore guarantee the "function" of the enterprise, which in a living organism is represented by work and heat. In the case of the organism work is understood as a motor activity in relation to the environment, not as the productive activity which it is implied to be in the case of the enterprise. The enterprise transforms mass into an "informer". It therefore needs raw materials. These are the basis for the product which is turned out by the activity of the enterprise. But in order to work the machines, the enterprise needs energy too. What is it that the enterprise adds between the raw materials and the finished product? It adds information. Where does this come from? Circulating information is of two kinds. One is that which training has internalised within the nervous systems of the individuals who make up the workforce (we have already seen that its degree of abstraction controls the hierarchical level attained by the individuals). The other is that which is introduced into the machines by the scientific engineers whose imagination has invented and "programmed" them. This occupational and mechanical information is the fundamental basis of productivity. In fact it is not the quantity of raw materials or energy used that essentially controls production, but the structure of the enterprise. An increase in the occupational capacities of the workforce, in the number of individuals or particularly in the number and effectiveness of the machines will result in an increase in production. By the same token one might feel tempted to say that there is another kind of information which controls the activity of the enterprise: information about the market. In reality, however, this information does not seem to be a factor in productive activity, but rather to represent the feedback loop of the output on to the factors of production, the set of factors we have just been looking at. In fact, both the value of the factors I have listed and the scale of the structure of the enterprise are controlled by the way in which the enterprise comes to be used by the local human sets on a regional, national or international scale.

Now we come to what seems to be the fundamental problem, the finality towards which the enterprise is directed.

(1) If this finality is the maintenance of its structure, without

any desire for expansion, domination or monopolistic growth (as was the case in the last century for numerous family enterprises), it is the structure itself of the enterprise that will regulate the level of production, and this will regulate the value of the factors necessary to reach this level of production. What we have here is an effector regulated in a steady state, with negative feedback.

(2) If its finality is to supply an industry which encompasses the production of the enterprise, it is this encompassing industry that will ensure the flow of commodities and thereby regulate the level of production, by means of a servomechanism of information coming from outside the system. The system will therefore be regulated, bearing in mind the possibilities of production that result from the structure of the enterprise. The enterprise can adopt a non-expansionist finality, with other enterprises of the same type ensuring joint production. What we have here is again a system regulated in a steady state, but the control is now situated outside it. This is how it should be in a system which is planned. Competition will result simply from variations in the effectiveness of the enterprises. Consequently the enterprise with the greatest production, the best quality and the lowest factor values will be sure to survive, or even to expand (which will raise other problems). But in this case the enterprise's effectiveness will depend upon its actual structure. Hence the advantage of a workforce that is technically very advanced, and of materials that are highly productive, novel or even revolutionary.

(3) Finally, if its finality is not production but profit, then it is all quite different. To increase profit, more must be produced. But the flow of commodities must be guaranteed: needs must be created. Hence advertising and the search for new outlets and international markets, with consequences which we have already examined.

With profit as the finality it is necessary to lower the value of the factors of production, while at the same time increasing its scale in such a way as to maximise the margin between receipts and costs. But it is rather difficult to pare down on raw materials and energy, and so attempts are made to extort them from the non-industrialised countries at the best possible prices: hence, at the national level, imperialism. On the other hand, productive investments are increased while wages are reduced. But above all, since the finality is no longer to participate in the finality of the set, and since each enterprise must obey these laws to survive, the result is what we know as "the consumer society", in which production is no longer important in itself and is no longer aimed at improving the conditions of life of individual human beings, but only at the profit to which it gives rise. If certain enterprises have complementary activities, the ensuing set may constitute a state within a state — or even a state within states, where there are multinational connections of accumulated capital. The pursuit of profit is simply the pursuit of domination, at all levels. In such a system, the individual is reduced to the role he plays in production, and necessarily takes his place as a member of an occupational hierarchy. His sole incentive is promotion in this hierarchy, either by virtue of seniority or by the acquisition of occupational information of a more abstract character. This is the purpose of re-training and permanent education. As detail labour is rather depressing, the system tries to make the worker forget its inconveniences by rewarding him with "leisure" and a predigested pseudo-culture which is programmed from start to finish, and which above all avoids any challenging questions. It is not only labour which is divided into detailed bits; this is now also the situation with human personalities, which cannot see whether life has any other meaning than that of improving one's "standing", since one's basic needs are taken care of (which is in fact far from always being the case, even in societies of high technical development). So the human personality is divided into two broad sectors: the economic sector, which also includes family relations and friendship, and the pseudo-cultural sector, which is a reduction to the non-scientific — for science is kept for innovation, i.e. for economic productivity. Culture carries on as if it were not for sale and not productive; in reality, while it is true that culture is a-scientific, it is also true that it sells very well and is an aid to selling. And sometimes, but increasingly rarely if the truth be told, economic man and man at play are joined by a particle of religious or political man (they amount to much the same thing). If the first two do not succeed in quietening the fundamental anxieties, and if these anxieties have not already been completely concealed by the socio-cultural automatisms, religion and politics are there to supply an activity and the pursuit of a gratification which is otherwise lacking.

Alongside the hierarchical relations in productive activity, the individual's relationships with his surroundings create habits in his nervous system which make him dependent upon others and others dependent upon him. Unfortunately if interference occurs in one direction only, a syndrome may arise which is analogous to that which develops when an individual is deprived of a drug on which he has become dependent. Here we have a case of real suffering, although no one can be called responsible for it; history is not like real estate — it does without brokers. In the end, for the individual as for the group, the thing which is to be feared most is the closure of the structure within too narrow a system. From this it follows that if we hope one day to achieve a stable social structure in which the individual can find his satisfaction, this structure will have to provide for an accelerated "turnover" in its elements, so that the individual is no longer subject to excessively stable relationships which may cause suffering when they are broken. Not only hierarchical systems relating to occupation but all social structures — starting with the family — will have to be thought out anew. Gratification must come on the basis of relationships among an extremely large number of elements in the social framework, and must never be institutionalised. If the search for constant repetition of acts of gratification forms part of the basis of neuro-physiology, as it does of toxicology, then at least we should get to know its laws and avoid addiction to poisons of whatever kind — whether chemical, social or cultural.

10

The power of the functional classes

It is possible to imagine a structure in which "power" is not a property reserved for certain people: a structure without power, or more precisely a structure in which the inter-relationship of elements is such that no one element dominates another or is dominated by a third, and thus a structure without a hierarchy of values. The important thing for human evolution is the constant transformation of structures, on the foundation of a basic outline which is linked to the specificity of the structure of the human brain. For example, a human set could agree to put into practice an original form of social organisation (a structure) proposed by a man or a group of men without making this man or group of men the "boss" of the organisation and without giving him any "power". If the organisation of human societies is inevitable because it is indispensable to their functioning, "the power" of one man or group of men must be banned.

This utopian conception demands that power be generalised to all the individuals in a society, and that their decisions should be identical. This is only possible if information is made uniform and if the motivations and interests are themselves identical. *Uniform information* is conceivable only in an authoritarian system where automatisms of thought are created from childhood, in accordance with a grid which is regarded as universal and definitive, and which no longer allows adult human beings to exercise their orbito-frontal lobes, i.e. that part of their brains which exercises the imagination. Obviously this is the complete opposite of what is desirable for human evolution. In this sense, "leftism" is as indispensable to the evolution of human society as conformism (whether of the right or of the left) is to its sclerosis. Grids always make prisoners of somebody or something, whether they occur in cells or in ideas. Uniform information always derives from a central power, and for this very reason it is incompatible with the generalisation of power. One can bring about uniform information by welding a "grid" into the nervous system at an early stage. From then on, only what belongs to this grid will be perceived as information. This is precisely what a "language" is, although information can be translated from one language into another. But for a person who has only one grid, the information transmitted through another grid or language will be unintelligible. Consequently everything becomes simple. Stereotyped action is merely the expression of a thought imprisoned by its own engrammed schemas, uniform formulas and routine concepts; these are expressed in a terminology which is never questioned, and no attempt is ever made to understand precisely what its intuitive or historical content is.

It is a utopian conception, which would also require that the motivations and interests of everyone should be the same. This uniformity is sometimes put into practice when one nation goes to war against another, and the dominant classes succeed in making the dominated classes believe that one single interest is at stake, the fatherland. We should note carefully that this unity is always created by the belief that there is a national "power" to be defended against the inroads of a foreign "power". In this case political parties, social classes, ideologies and religious beliefs unite in rebuffing what is known as "the common enemy", although this common enemy contains the same parties, classes, ideologies and religious beliefs, and a "horizontal opening" between them is just as conceivable as a "vertical closure" at the frontier of the national framework.

It cannot be accepted that information should be made uniform within a national group (religious group, group of citizens, regional group, enterprise etc.), nor that the activity of the set should be governed by a class interest, a social or functional interest (corporatism) etc. What we must seek, therefore, are ways of generalising and diversifying information on the one hand, and on the other hand a finality which is internal to the system and is linked to its structure rather than its thermodynamics. In other words, we must try to separate inter-human, sociological, informative relations from the iron yoke of the thermodynamic relations of production.

I have already tried to move away from the automatisms of thought associated with words such as "workers", "social classes" etc. An expression such as "manual and intellectual workers" is an encumbrance for orthdox marxism; it fails to distinguish between thermodynamics and information, and as a consequence it blindly preserves the hierarchies of value, wages and power. Where power is concerned, orthodox marxism repeatedly states that the class exercising it should change (power to the working people, to the working class); the interests of the workers are spoken of as if,

175

come the day when there are only workers (manual and intellectual, of course), there will no longer be any contradictions or class antagonisms. State ownership of the means of production will suffice to bring about the abolition of social classes, the full flowering of the humanity and the ability of the individual ultimately to become master of his own destiny by means of democratic planning.

Political power based on the indispensability of functional classes cannot relapse into new forms of domination and new hierarchies of social groups, if the information concerning the vertical and horizontal structures is broadly diffused and if the finalities of these structures are clearly laid down — that is to say, if the finality which expresses the action of the largest set (the social set we are currently dealing with is the nation) is given a clear definition. But a programme of this kind does not simply amount to making a list of improvements to be carried out in the various spheres of individual and community life. Nor does it mean using vague general phrases about the full flowering of humanity, equal opportunity, improvements in the quality of life, the individual being master of his own destiny and so on.

Such a programme must first of all define what will take the place of economic expansion: the motivation of each individual, of each element in the system, will depend upon this. With expansion as a means of achieving the finality of domination, the motivation of individuals and social groups can only be to ascend the hierarchical ladder of domination. In order to satisfy man's congenital narcissism — the need to be loved and admired which every individual has from birth to death and which constitutes the emotional basis of his behaviour as soon as his basic energy needs have been secured — would it not be possible to encourage, not power and hierarchical domination, but creativity and imagination? Instead of motivating the child to be first in class and then to find a well-paid and "respectable" position, or motivating the individual to rise in the occupational hierarchy with an ideal of social advancement which becomes real when he possesses consumer goods, would it not be possible to motivate him to imagine new structures which have not so far been thought of, in whatever discipline this may be? Would it not be possible to encourage this kind of creativity, not by material gain and not by occupational or political hierarchical power, but simply by public recognition of this creativity? If any form of individual property or ownership is to be retained (although the notion of property itself is debatable in this context) it is the ownership of imaginative creation, since there is only one truly

176

human characteristic and that is to know how to create information.

While it is absolutely essential to avoid all personal "power", it is at the same time possible to recognise an individual "role", as long as this "role" is not accompanied by any power. Power must be kept for the functional group. The political power of the functional classes (and not exclusively of occupational ones) depends on their indispensability, not on the degree of abstraction in the occupational information furnished by training. This power of the functional classes will not lead to corporatism, because it will be directed towards *complementarity* (as defined in the theory of sets), not towards antagonism. Antagonism leads to the disappearance or subordination of one of the two parties in the conflict; it ends in domination. This is an inevitable result of the fact that attention and action is focused on an incomplete structure, which is seen in terms of limited sets and not in terms of the largest possible set.

A power of this kind, belonging to the functional classes within structures which are increasingly complex both vertically and horizontally, clearly demands the extensive diffusion of information about the dynamics or "cybernetics" of these socio-economic structures; but it also calls for the diffusion of information about the thermodynamics underlying them — i.e. about the economy or, if you like, about the manner in which energy and raw materials, taken from the environment and transformed and then redistributed by man, flow through the structures.

There is today a lot of talk about "participation". But this should not be participation in profits, which in a society based upon expansion and hierarchical domination will be restricted to participation in the profits of an *enterprise* and will thus recreate intergroup competition and facilitate the maintenance of occupational hierarchies, by holding out the expectation of equality of consumption. This kind of participation will lead to maintenance of the institutionalised structures of domination, and this in fact is its real, though unavowed, goal. The surplus value created by human labour exists in all socio-economic systems, whatever its degree of abstraction (and generally the greater its abstraction the greater its scale, since today information feeds machines which partly replace the human being's energy-labour). In capitalist systems it is to a considerable extent reinvested, in order to ensure the continued domination of the dominant; part of it is redistributed to secure a growth in the purchasing power of the masses, whose consumption is directed in such a way that profit may increase and thereby allow the domination to be maintained. In the socialist

177

countries, surplus value becomes the property of the state and thus in principle of the collectivity. But it is not the collectivity which decides how it is used but people in a position of domination of another kind: bureaucrats and technocrats who forbid the dissemination of information about any other socio-economic and political grid that might call into question their domination. The institutionalisation of the rules, prejudices and mental automatisms which are qualifications for membership of the bourgeoisie in the capitalist countries has been replaced in the so-called socialist countries by the institutionalisation of rules, prejudices and mental automatisms which qualify one for membership of the party. In neither case have the dominant and the dominated disappeared; the hierarchies are still there. Thus participation in the redistribution of the surplus value of enterprises, which is made so much of in certain capitalist countries, is a means of chaining the worker not only to his enterprise but to a whole system of life — production for production's sake — whose dangers I have already pointed out. Seen in this way, participation works in a direction which is completely opposite to that which I have indicated as the desirable one. It encourages restrictive corporatism within an enterprise; it is oblivious of vertical and horizontal openings; its sole motive is the growth of ownership and consumption; it is an incitement to intergroup domination and monopolism; it encourages ignorance of the power of the functional classes; it divides instead of uniting; it creates oppositions instead of composing complexes. Participation in profits, in the sense of the harmonious redistribution of surplus value, has in principle already been carried out by contemporary socialism, at the level of the state organisation. But the individual is totally ignorant of what happens to the surplus value that he creates and how it is used. He is quite powerless to take any action over its mode of utilisation because he has no political power and no information about the national and international structures and their thermodynamic activities, because political power is "held in trust" by a limited group of individuals, and because even the strike, the one possible means of expressing political power, is forbidden. Thus the state's participation in and redistribution of surplus value remain subordinate to the occupational and in particular the bureaucratic hierarchies. No new satisfaction or "nourishing" motivation ensues. An atmosphere of resigned sadness develops, which is not brightened by the recitation of stereotyped phrases learned by rote, nor by automatisms of thought which are inculcated from infancy and conform to a marxist grid that differs in interpretation from one part of the planet to another.

Participation in decision-making on questions which concern the survival of the enterprise is usually called "self-management". But from everything that we have discussed so far, it is easy to see that this type of self-management is conceivable only on two basic conditions: (1) knowledge of the existence of the openings in the enterprise's closed system, both in a vertical and in a horizontal direction, and knowledge of the mechanisms of these openings; (2) the abolition of occupational hierarchies and their replacement by the political power of the functional classes. This political power must itself be nourished by a knowledge of the vertical and horizontal openings, and by a knowledge of the levels of organisation relating to energy (the economic level) and to information (the organisational level). Without this, self-management will once again end up in the closed corporatism of the enterprise, or in a restrictive bureaucratisation which reproduces the system of occupational or trade-union hierarchies.

And so, to return to the ideas which we have already discussed, self-management only has a chance of being effective if generalised information is widely diffused, and if every individual has the time needed to receive this information and to integrate it with what he already knows.

Trade unions and parties

Thus we come to the question which has been debated thousands of times concerning the respective roles of trade unions and parties. When one reads that the role of trade unions "ought to be limited to the defence of the workers' interests", one sees that in people's minds (just as in their conditioned behaviour) there is a solidly anchored belief that material well-being is distinct from domination, and that it is possible to be perfectly happy without political power. Moreover, the palaeocephalic motivation which forms such an opinion can immediately be guessed at: it is so that the parties — that is to say their political leaders — can maintain their domination. In order to open a closed system, it is necessary to supply it with both vertical and horizontal relations. We shall examine later the consequences of ignoring this fact, and the idea of structures and information on the one hand and thermodynamic support on the other.

What are the "interests" of the workers? Are they purely thermodynamic, related to nourishment and to the actual conditions

179

of work, or are they also organisational — that is to say, informational — and related to the ability to act upon the structures? If the latter is the case, then we know that a structure cannot be limited to the level of organisation of the enterprise, and that vertical structures always lead to the largest set; this unfortunately is still considered to be the national set, when in reality the largest set is a planetary one.

Clearly, trade unions till now have generally been restricted to bringing together the functional classes according to criteria determined by the occupational hierarchies — workers, staff, bosses etc. — within the various fields. So what we get is a horizontal opening with a vertical closure. This vertical closure then rules out the possibility of trade unions expressing an opinion about structures, since this would necessarily involve the largest set in both directions, vertically and horizontally, as well as its finality. Consequently the trade unions have had to restrict themselves to thermodynamics, to questions of purchasing power and working conditions. They have used their "power" only in this restricted sense. They have allowed the indispensability of certain functional classes to be felt, but in order to secure greater economic power rather than greater structuring power.

The political party, on the other hand, is based on a vertical opening. It unites quite varied functional classes within itself. It cannot use the strike weapon to express its power without having recourse to the co-operation of the trade unions, which because they are not organised vertically but horizontally cannot effectively transform structures, particularly vertical structures. Systems of occupational hierarchies, which are based on the degree of abstraction in specialised information and supply the current sociological grid, must always rule out effective participation by the trade unions in the organisation of informational social structures and in the establishment of vertical structures. They must likewise prevent the power of the political parties from being extended horizontally to the functional classes — unless this is done by means of shameless demagogy and promises made to each functional class about meeting their thermodynamic demands, when in fact the role of a political party can only be structural and in the vertical direction. A political party can precipitate a political crisis, but not a strike.

Power can only be the property of a functional class, and is based on this class's indispensability within a structure which is not hierarchical but complex. Trade unions must therefore combine their horizontal opening with a vertical opening; one may or may not call this politics, but at all events it is connected with the organisation of vertical structures. Likewise, political parties must adapt their proposed vertical structures to the existing horizontal structures. But to propose a vertical structure in which hierarchies of value have disappeared cannot imply maintaining the hierarchies of value within the party organisation. In reality, the political game has become an occupation, which is founded not upon creativity but upon the most biologically primitive type of domination, because the individual's occupational activity prevents him from informing himself in a general way about the laws and dynamics of structures. Certain individuals therefore become specialists in the acquisition of this knowledge, and then take up positions which are hierarchical in relation to the uninformed mass. Their domination gives them satisfaction, and they cannot imagine quitting the ideological framework of the party, for this would mean losing their domination. This is one of the main reasons for the ideological sclerosis of political parties.

11

Occupational information and generalised information

The train of our argument has led us to single out an "information-structure" which consists in the giving of form to the cell, the organs, the human body, the social body and the species. This information-structure is constituted by a set of relationships which one is tempted to call invariant. But although they are immersed in their inanimate and animate surroundings, the information-structures do change, by means of "exchanges" with these surroundings; the time-scale, however, is that of the evolution of species or of the individual over his whole lifespan, not that of immediate response to the environment. For example, the nervous structure of the individual changes between birth and death, because of memorised experience and ageing — but he still belongs to the same species. He conserves the "structure" of the species, its programme, and the genetic information which was originally supplied to him.

We have distinguished this from "circulating information", which resembles what the telecommunications engineers study: it is what they try to protect from noise and interference, in order to preserve the semantic content. To put it another way, whereas the biologist is chiefly interested in the signifier, in the way in which the letters in a telegram are associated in monemes, phonemes, words etc., the engineer is interested in the telegram's quantity of information and in not losing any.

There are no value hierarchies among the elements (letters, monemes etc.) which are organised in the signifier of a message. There are only differentiated functions. But a message acts only by means of the circulating information of which it is the bearer. A living structure, an information-structure, has no message to transmit to the biologist studying it: it simply exists.

On the other hand in order to exist as a complex structure with levels of organisation, and in order to act upon the environment with the purpose of maintaining this structure as a function of the variations in this environment, "circulating information" or messages must be transmitted to all the elements in this structure. This

182

is what is done by hormones (they are "chemical messengers") and by the chemical mediators of the nerve impulse; they are capable of giving a rapid warning to the entire set of cells in an organism, or perhaps to a particular cellular group performing certain functions, about the necessary metabolic effort (involving not only energy but "occupational" activity, since it varies according to the specificity of the function performed) for maintaining the economy of the organic set. Conversely, as a result of the stimulation of certain cellular structures that are sensitive to variations in the internal environment which themselves result from the functional activity of different cells, tissues and organs, this organic set warns the nervous system of its "desires" and "needs". The nervous system brings into play the organs which will permit motor activity upon the environment, and this activity must strive to satisfy these desires and needs.

If we consider this now at the level of organisation of societies, again we find an information-structure which regulates inter-individual relationships between the individuals who make up the society. This information-structure depends upon its finality, for this is what will determine the circulating information.

I have already indicated repeatedly that circulating information may assume the form either of specialised, occupational information or of generalised information. Let us now go a little further into the forms which are assumed by circulating information, since it plays such a fundamental part in the organisation of the information-structure.

I shall not deal at length with specific, occupational information. All teaching, from nursery school to university, provides precisely this. The reason is that the hierarchies of domination are based on this kind of information, because it permits the accumulation of capital through the mediation of massive commodity production. This definition may appear to overlook so-called cultural information, which is valued so highly because of its "ennobling" effects on people. But the content of this "cultural" information is suspect from the outset, because it appears as a domain apart, i.e. as a closed system. How can there conceivably be two domains so different as those of work and "culture", within the realm of human knowledge? "Access to culture" in fact means access to the passwords of a secret, closed society, a cultural Rotary Club (as opposed to an economic one) in which the individual is classified according to a hierarchical scale of gratification.

183

He is not required to establish personal syntheses in the various fields of this pseudo-culture; he is simply required to be able to conduct the kind of conversation which expresses his level of education. More importantly, this commodity and production-minded society feels deep inside itself an obscure resentment towards the limitations of its commercial activity. Culture is, in fact, chiefly what is useless. Art is good for nothing: you can get along quite well without it. It is an extra, something you treat yourself to once you've got a refrigerator, a car and a cottage in the country. With a bit of skill, you could use it as a commodity and improve your capital. In a society of shopkeepers and speculators, it is the only thing you can deal in with a clear conscience and a disinterested air. Art is a souvenir of the old ruling class, of the world of salons, titles and ostrich feathers which hangs about even now in the unbreathable air of our modern cities. The *"bourgeois gentilhomme"* is far from dead, in fact he continues to reproduce to remarkable effect. "Cultural" information is separated from anything living: it is a trinket offered to people as the illusion that they are improving their hierarchical situation and thus achieving gratification. In actual fact, cultural information is not useless. But in order to be interpreted or even understood, it needs a socio-biological code to restore its meaning and allow its message to be included in the whole of human behaviour in the social and historical situation. The creator's motivations. the nature of his distress, his fruitless search for pleasure, his socio-cultural automatisms, the material used by his imagination, in short his unconscious: all this is engraved upon the social history of any given epoch. Similarly the way in which he is understood, the way in which his work is decoded for himself or for the group by those who receive it, is not a phenomenon that can be separated from the biology of behaviour. It is chiefly a bio-sociological phenomenon, and it belongs to a system, an information-structure and a hierarchy of domination. Volumes could be written on this aspect of the works which the creators have given us, which string words together like pearls on a necklace and develop a logical argument to interpret the logic of the dream, which in turn knows only the complex biochemistry of the functioning of the nervous system. One could call this culture. Probably it is simply a supplementary training, whose main advantage consists in the fact that it does not need to be experimentally tested and therefore leaves a wide latitude for the expression of "originality", something which the technical automatisms of today render somewhat rare. Since this originality of opinion has no consequences for the individual production process and is hardly likely to transform the social structure on its own, the hierarchies have every reason to develop this mechanism of gratification amongst the masses, for it does not challenge their domination. It can even be used to strengthen it.

When I speak of "generalised information" it is not to this type of cultural information that I am referring. What, then, is this generalised information if it is not occupational information and at the same time does not come under the narrow heading of conventional "culture"? We can outline its methods and objects as follows.

Methods

Our nervous system is born immature. The inter-neuron connections are enriched in the course of the first months and years of life outside the womb. Then the coding of memorised automatisms fixes our nervous system in a structure which will be the basis of all our value judgements and from which it will be practically impossible thereafter to disconnect ourselves. Once again, what we have here is a closed structure which will be deaf to all information for which a place has not been prepared in advance; it will be incapable of opening on to larger neuronal sets, and will be fixed in its conceptual automatisms. So the first years of life will be of crucial importance — not to stack the memory with habits, but to prepare the kind of structure of neuronal interrelations whose opening thereafter can come about in a practically infinite number of ways.

The *theory of sets* (which it would be more appropriate in this case to call the theory of relationships) is an indispensable basis to this culture, not only because it opens the way to mathematics, from its most simple to its most complex forms, but particularly because it enables the world to be apprehended and organised in a coherent and logical way: this applies as much to the world of physics as to that of biology and as much to the world of matter as to that of the concepts which derive from it. The theory of sets is not a game like the rules of bridge, which the child would be taught and which he would then either forget (if he pursued, for example, a literary career) or use occupationally in the course of a scientific career. The child should get into the habit of using this as a language, for it is a kind of logic, that is to say "the" modern logic. To deal with the world of our present epoch without this internal language and logic is to deal with it in a state of hallucination. It is a marvellous vehicle for the exchange of precise information between human beings, and can help to make good to some extent the dangerous lack of precision in affective and cultural language; it is a means of avoiding the mixing up of levels of organisation. Finally, it is an open system which proceeds from the simple to the complex and provides the individual with a conceptual tool that is capable of constant enrichment. It opens up to him the world of structures, the world of information — a world to which we have been oblivious or at least one which has been isolated from the other world by having the label of "mind" applied to it, because it organises our very nervous system. Until now the "mind" has been no more than a jumble of value judgements.

This initiation into the theory of sets must be complemented with the rudiments of cybernetics, which will put life into these structures and allow them to evolve over time. Again, here is a marvellous instrument for knowing the animate as well as the inanimate world. Cybernetics makes it possible to pose a problem correctly: who does what, how, and why? It enables us to set up material and conceptual models and to check their functioning. It compels us to avoid confusing the effector with factors, effects and feedback. It compels us to look for the greatest number of these in a system and to escape from the linear causality and the three Aristotelian principles by which we still live. Above all it forces us to reveal the levels of organisation and the intervention of the servomechanisms. It leads to the *study of systems*, both open and closed.

The object of this "structuralist education" (to give it a name) is most certainly not to automatise virgin brains, but on the contrary to open them up to the world about them and to the world which lives within them. It will add some order and clarity to the infinite complexity of things, beings and concepts, the jungle of the unconscious, the barriers imposed by taboos, and the implacable control exercised by the automatisms. It will enable human beings to act like human beings and not like talking chimpanzees. It is easier to communicate than the multiplication table. It hardly involves the memory. It demands "comprehension", in the etymological meaning of the word. Above all else, it enables us effortlessly to keep in touch with the sensory world. It can attach itself to life at any moment, penetrate it and become part of it: not simply occupational life but life *tout court* — the life we meet with every day, what we read about in the paper when we come home in the evening, the family problems or social problems which have to be solved,

186

as well as the life of international relations. It shakes up values, all values, from the most obvious to the most debatable. It tirelessly questions everything. It incites revolt against prejudices, outworn concepts, original truths, "essences", admirable certitudes, morals, ethics, and words — any words which don't lead to writing a poem and tearing it up once it has been written.

In reality, it simply extends the specific information-structure of the human brain. It organises the infinitely complex network of neurons into an open structure capable of accepting and giving order to everything. It becomes circulating information as soon as it penetrates this network. It accordingly becomes information-structure too, but a structure which is capable of accepting everything without disorder, by pursuing at a conceptual level the negentropy which has been introduced by living systems into the world of matter. It may be said that what this "structuralist education" does is no different from what all education does. But unfortunately, education has so far been a closed system, introducing into the nervous system "elements" which in the absence of an open organising structure necessarily lead to value judgements and socio-cultural automatisms, that is to say to closed structures and the conflicts they generate. People have written[48] of the informational aggression to which modern man is exposed; this is probably due to the fact that man has not yet learned how to organise his information on the same scale as he receives it. He piles it pell-mell into his nervous system, where the only things which surface are those bits of information that fit his drives and automatisms. What doesn't fit he buries or doesn't even hear, for this would then be a source of distress and conflict. Nevertheless, this does not enable him to avoid conflicts between the closed structures which reside in him. Pre-existing grids, imperfect and likewise closed though they may be, sometimes help him to survive this distress; but they only plunge him subsequently into blind and homicidal sectarianism. It is unusual to find symptoms of informational aggression in someone who is conditioned by, for example, a marxist or a psychoanalytic grid. All information is immediately fed into the grid and decoded. It emerges in the explanatory form of the class struggle or the Oedipus complex and no longer presents any problem about what action is to be programmed in response to it. If it cannot be introduced into this schema then it is simply meaningless or uninformative; it is not heard and thus doesn't pose any problems, since it cannot influence action which has been completely programmed in advance. Even if the new information so insistently

187

presents itself as to release a potential for action at the level of the neuronal circuits, it will provide material for that admirable thing known as "analysis" (a term used in both grids), which is a logical discourse of linear causality and is generally unconscious of the unconscious by which it is guided. So informational aggression only exists for people who have no grid at all, or at least only possess a set of value judgements and socio-cultural automatisms, and who cannot be told that the latter have been invented to preserve the hierarchical structure of domination in a given society and in a given epoch. These value judgements and automatisms therefore constantly contradict the observed facts because the key to their interpretation is lacking, having been carefully hidden away. Incoherent, contradictory and increasing information thus cannot open the way to an act of gratification, particularly since in the absence of objective observation each user of a grid claiming to describe the facts propagates (each according to his own grid) different opinions and judgements on one and the same observed phenomenon. Then why, you may say, add another grid to those which already exist? Because the new grid comprises the old ones, plus a certain number of fundamental elements which they do not contain. But I am quite convinced that this new grid will only be a temporary one; and I hope that it will not pick up en route the sort of automatisms that could make it an instrument as rigid as those of Aristotle, Freud or Marx. However, this grid tends to be protected by its openness, for it is capable of questioning its own existence as well as the temporary value and transitory efficacity of grids in general. Relativity has not undermined the laws of gravitation, it has encompassed them.

This structuralist education leads to a distinction being drawn between information, mass and energy, this underlying trinity from which we shall certainly not be able to escape for a long time. We can introduce new elements into this structure in considerable numbers without fear of indigestion or informational "stress". They will logically take the places reserved for them in the structure. The amount of effort to be exercised in the form of memory will only be slight, or to be more precise the phenomenon of memorising will be greatly simplified. The disorder of classical "culture" will be succeeded by the order of a puzzle which one has only to complete. But the metaphor of the puzzle is imperfect: the puzzle is a closed structure, whereas memorised facts will be able to enter into an open structure to an infinite extent, the sole and constantly changing limit being that of our knowledge. The greater part of the time lost in a human life will be recovered, for this time is lost because of the sterile games between sub-systems, incomplete sets and scattered elements. The child will learn to never let himself be enclosed in an antagonism, or rather to use his imagination to discover the new structure which will enable him to get out of it. There will be a *fuite en avant* which is based upon the constant enrichment of knowledge, and made possible by the opening of nervous and hence conceptual structures.

Objects

The number of objects is infinite, since what we are talking about are objects of knowledge. But at least we can envisage a general framework and, bearing in mind the methods we have just been discussing, a way in which to organise this knowledge.

Let us deal first of all with knowledge about the physical world. Perhaps we should not follow a historical sequence but attempt to show how, starting with the basic principles of thermodynamics and kinetic theory, the electronic structure of matter enables us to proceed from physics to chemistry. We can also show how enzymatic reaction allows us to proceed from the chemistry of minerals to the chemistry of biology, which are sustained by the same laws. We can see how the laws of evolution have led from the sun to the human being[44] in terms both of energy and of information. It is quite superfluous to add Shannon's formula or the law of entropy. One of my children, then aged eight, said to me once "so we're eating sun", showing me that he had understood perfectly, in a very simple formulation, the informational evolution of the biosphere starting with solar entropy. We need to place the individual in his cosmic set from a very early stage, to make him a participant in it instead of teaching him to isolate himself from it. We need to teach him also from a very early stage that he is a particular small animal which talks. We need to teach him how his brain and that of other human beings has been organised over the course of the many millenia which separate the first living forms from the first men. We need to teach him how the human being organises himself from the moment of his birth, what factors make him act. There are very many ways of saying all this, ranging from the extremely simple to the extremely complex. There are ways suitable for all ages, from seven to seventy-seven. With such a grid at his disposal, an adolescent might be able to decode all the comics he could find and decipher their unconscious message, the sociolo-

189

gical expression which they represent, the cultural conformism which they transmit and the automatisms of judgement which they are able to cause, even if their editors are unconscious of them. From the moment he is capable of finding out, beneath the discourse of the other, the purpose which guides him, it will only remain for him (and certainly this is the most difficult thing) to find out the purpose which unconsciously shapes his own discourse. He will be ready to enter social life.

Social life at the outset will for him be linked to his occupational function. We can be quite sure that contemporary societies will maintain for as long as possible the kind of education that offers a "good start in life". As long as the human being is no more than a producer of commodities, he will have to find himself a place in the production process, either at a later stage if he has first to equip himself with the abstract knowledge of occupational information, or early on if he only uses his hands. But in either case he will be given two wishes. The first is that whatever work he does he may find more than a hierarchical interest in it: for this to be the case, it must not be too fragmented. The second and more important is that the methods of thought by which his nervous system has been shaped will make him discover something different, another interest in life beyond his purely technical, occupational interest. To escape from this today, all he has at his disposal is the framework of the family (bourgeois or otherwise), with its Sunday get-togethers, television and weekend car-excursions, or playing the horses, sport, "culture" etc. One may ask whether all societies have not always favoured this kind of diversion, using conscious or unconscious propaganda to focus interest on this trivial framework to prevent their existence from being called in question. It is desirable that every human being should feel himself concerned with all the social structures, at levels of complexity ranging from that of the individual to that of the species. As a social animal who cannot exist without others, man should be interested in relationships (or the lack of them) with those who live on the same storey as himself, but also with all those who make up the functional classes to which he himself belongs. These functional classes do not only consist of those which perform an occupational function. They are all the classes which spring from the formation of human groups of varying duration and stability, with a specific finality, a particular function in the social set. In other words, it is desirable that he should be conscious of the existence of other relations apart from the productive relations — those which Freud studied, for example.

190

I would in particular pick out those which structuralist and dynamic notions of modern biology have brought to light concerning human behaviour in the social and historical situation. After that he will not be able to limit his interest to the restricted group of human beings who people his own immediate and current environmental niche. He will know that every human group which does not open on to a larger set, every closed system, is an antagonistic system destined either to disappear because of the domination of another closed system or to become dominant itself, that is to say to become part of a hierarchical encystment. The term "opening", which is very much in vogue today, does not imply the integration of others into its own structure (which in that case would no longer be open). Neither does it imply the dissolution of this structure in another set. It implies the formation of a new set, whose finality changes when the two sub-sets become complementary in order to define a new level of organisation. The first two interpretations, on the other hand, lead only to the preservation of a single finality which is that of the dominant set, with a simple increase in the number of elements in this set.

All this is only possible if the circulating information truly circulates, and if each element possesses a panoramic view of the problem posed (this applies just as much at the level of socio-occupational activity as at that of the social structure). For this, the acquisition of information must clearly not be the divine right of certain dominant elements in a structure of domination, for in this event the only information to make any impression on them will be that which enables them to perpetuate their domination and preserve the social structure in which it is embedded. We have already noted that each element in the social set must be able to obtain, decode and interpret information; this presupposes that there is an effective grid for so doing, comprising other grids; without these, it can only be a closed (and hence antagonistic) system. Each element in the social set must therefore enjoy free time, as at the beginning of the neolithic period, so that it can inform itself.

Circulating information must thus make available everything which relates to the information-structure, or in other words the anatomy and physiology of the social body. This must not be a process which comes from the top down, with the top putting out only pre-cooked information or filtering it so that it encourages acceptance by the base of the existing hierarchical structure. Not only must censorship disappear but also the self-censorship of those

191

who want to secure their promotion in the hierarchical system, whether consciously or (as is more often the case) unconsciously. It is usually unconscious, for the hierarchical gratification involved in submission to what is euphemistically called the "national interest", and the conditioning it produces, simply express the determinism which guides the unconscious behaviour of the human nervous system in its pursuit of domination. Information must circulate in its contradictory aspects; this is possible once there is no longer a hierarchical structure of domination to be protected. Circulating information should supply an inter-individual concept combinatorial similar to the gene combinatorial, instead of encouraging a sort of eugenics of concepts, a conceptual racism which runs contrary to all biological evolution. We should simply note that once the set has formulated or discussed a working hypothesis, it is only the base which is in a position to test it.

Within this continually developing social structure there flows a wave of energy which transforms the mass and the raw materials. This wave of energy is that of the solar photon, transformed first by the biosphere, then by man. It is the *energy* aspect of the problem, the *economic* aspect. No amount of technical economic trickery can avoid dealing with it. Where are the sources of energy? In the immediate ecological environment, in the community's territory? Or in the extra-territorial environment? The distribution of this energy, in its primitive form or transformed by the information added by man, and the distribution of the raw materials that man transforms by means of it (which are necessary for the maintenance of the social structure by the survival of the individuals composing it), can be envisaged either in the closed system of the nation, or in the closed systems of transnational interest-groups (this is the main cause of wars), or in an eventual structure covering the entire set of the species. Even if this latter eventuality is not immediately realisable, an opening towards the largest possible set should at least be the backcloth for the distribution of energy and raw materials within human structures, as well as for the technical information which transforms mass and energy.

The problem of the *speed* and *effectiveness* of the decision-making process would necessarily take a long period of development to be resolved. As I have already said, in my view the decision-making process has never existed. The choice is always a pseudo-choice, expressing the drives and socio-cultural automatisms of the person who seems to be making the decision and who, because he is oblivious to these drives and automatisms, believes he is

192

a free agent. But even this is not really a problem. If I carelessly put my finger near a flame, I don't wait for all the disturbances that follow throughout my organism to "decide" to take my finger away from the flame. A rapid reflex action will put some distance between myself and the source of thermal aggression. In other cases my behaviour may be the result of previous experience, of the socio-cultural automatisms which I have acquired in some form or other: in this case I simply perpetuate a form of behaviour which has previously gratified me or saved me from displeasure. The child who has learnt to his cost that fire burns does not take a decision not to put his finger close to a flame. Quite simply he abstains from doing so. Any technician who has been properly instructed in the experience of his precursors by his technical training and by his own experience, does no more than apply the technical methods he has learnt. Those who think up new solutions are rare indeed. And even then the new solutions must be adapted to the environmental conditions, as well as being simply the result of the complex and unconscious mechanisms which control their behaviour.

This is the stage which modern societies have reached. It is sometimes valid for immediate purposes; but it is often ineffective, for meanwhile the whole environment has been transformed. The response which may be adequate for a particular set of events has little chance of remaining adequate for a new set, particularly if one takes into account the accelerating complexity of the modern world.

Finally, there is "imaginative guidance". This would be based on a finality which is accepted by the set and accompanied by as much scientific and conscious understanding as possible of the motivating factors which have determined its elaboration. It may be admitted that this can only become true once a finality for the "temporarily" national set has been adopted. In this case the rapid reflex response must conform with this finality: it may be only at a secondary stage that the circulating information will convey to all levels of organisation within the social organism the arguments and counter-arguments which have led to the forging of this reflex response. But the role of the imagination is chiefly to foresee what may happen and to demand the suggestion and discussion of possible solutions taken from the inter-individual conceptual pool before the event takes place.

One of the organism's basic motivations is its existential anxiety and lack of information about what is going to become of it. Contemporary social structures either do everything they can to draw a veil over this anxiety instead of using it to encourage the expression of ideas drawn from people's imagination, or conversely they encourage the diffusion of tendentious and anxiety-creating information in order to make a profit out of it. The "decision-making" process only appears to be such. Essentially it is conditioned; those who believe themselves to be trusted with the responsibility for making decisions are oblivious of the fact that they only fulfil it after the event. "Responsibility" is in fact no more than an appeal to their primitive drives, their socio-cultural automatisms and technical training. There is no invention involved: one reproduces what has already been seen to work. There is no creation, only reciprocal adjustments in a system which is trying to make itself eternal. Creation and imagination are only possible on the basis of a certain amount of experience, and this is not acquired by one element in a set but by the set of all the elements. The duty of exercising the imagination therefore falls upon the whole of the social organism. For this it is necessary that circulating information about the internal structure and its relations with the environment should actually circulate, and that it should not be enclosed within conceptual automatisms which only encourage the conservatism of the hierarchical structure of domination.

One can object that in a human organism the imaginative function, given the absence of a hierarchy of value, is not distributed in a general way. The objection seems to me fairly thin, for it is not the associative systems which exercise the imagination: all they do is associate the memorised elements without which they can do nothing. I have already pointed out that a new-born baby imagines nothing because as yet it has nothing in its memory, not even its own bodily outline. The function of imagining, which springs from the original processing of sensorial and interoceptive information, is certainly a global function. That which imagines is an image of the self evolving in time. Furthermore, every functional element in a social organism is endowed with imagination, if it has not been too rigidly automatised. So there is no reason why the imaginative function should be reserved to certain people who take advantage of it to establish their domination. And even if new solutions come only from certain people, these solutions should be discussed and accepted by the set, which ought therefore to be motivated and thoroughly informed in a contradictory way about the elements of the problem to be solved. Finally, there is no reason why those who propose new solutions to the problems posed should enjoy hierarchical domination and power. So the interindividual concept combinatorial which I am proposing is the analo-

194

gical expression, at a higher level of organisation in society, of the information combinatorial which springs on the one hand from the organism itself (summarising its state of physiological well-being) and on the other hand from the environment. The result of this is an act of gratification by the organism on the environment.

Circulating information therefore calls for a certain speed of diffusion, which audio-visual means can provide as long as sufficient time is allowed for reception and processing. It also calls for organs capable of decoding the information. That is why we have insisted upon methods whose sole aim is to enable the messages (which for modern man are particularly numerous) to be received and decoded properly. But for this, there must be no interposition of a filtering and deforming grid (i.e. the grid of hierarchical power) between emission and reception. The bits of information would be far less numerous if the message could be protected from the interference of value judgements; this interference occurs because the message passes through closed systems, which use it solely to defend their structure and pass on only a curtailed or deformed version. This is what the daily press does. Its various organs are read exclusively according to the affective criteria of the reader's hierarchical level and experience of gratification. The objectivity of information is well known to be a farce. One has only to look at the different ways in which a commonplace fact, such as a traffic accident, is assessed by a dozen or so people who were present. In telling the story, each person in reality expresses his own unconscious desires, value judgements and sociocultural automatisms. Only a diversity of information and a multiplicity of sources can possibly supply individuals with the material needed for imaginative and creative work. Indoctrination is always made easier by the ignorance of the person being informed; it presents him with only one aspect of things, tends to impose automatisms of thought and behaviour upon him, and conceals contrary opinions either by declaring them to be wrong or by presenting them in such a way that they immediately lose all coherence when faced with the solution prepared by the informer (whether this is an individual or an institution). Indoctrination is an expression of profound contempt for human beings. The human being is considered incapable of forming his own personal opinion because he is ignorant; this is indeed the case, but instead of his ignorance being made good by a supply of differing or contradicting opinions and information, he is deceived by being shown only one aspect of things. This is to consider him a subhuman creature; it displays a kind of racism.

195

The role of any power should be, not to "form" opinion, but to supply numerous elements of differentiated information, enabling each individual to question daily the basis for the permanence of this very power. This would mean suppressing all centralised power. It would mean providing each individual with a means of contributing his imagination to the always unfinished task of constructing human society.

The place for generalised information is not only in schools and universities, it is also in all human groups in the production process. Let us imagine that each person at work, from the age of seventeen to retirement at sixty-five, could enjoy two hours daily for acquiring information about problems which arise not only at the level of organisation of the enterprise where he works, but also at all higher levels of organisation encompassing this enterprise: municipal, urban, regional, national, international and planetary. Let's suppose that he does not tackle these problems solely with the aid of "rational analysis", but that he can also account for the part played in his discourse and in that of others by the unconscious, with its socio-cultural engrams, its prejudices and value judgements. Let us suppose that in this way he learns to distrust his own behaviour and that of others. Finally let us suppose that non-directed, multi-disciplinary and popularised information is supplied to him, stripped of its affective certitudes, dogmas or unwarranted affirmations, and presented "scientifically" in a way which is serious but can be readily absorbed. Finally let us suppose also that constantly contradictory information is supplied to him at the actual meeting-places of his enterprise as well as in the mass media and in the community where he lives. If all this were done, it is hard to imagine the wealth of remembered material his orbito- frontal cortex would then have at its disposal. This is perhaps utopian, but it is a utopia towards which we might at least strive to move.

The objection perhaps would be that this kind of generalised information can only be acquired over the course of a lifetime. This would probably only be true if the information were technical information, which would require everyone to be a polytechnician. If this was possible in paleolithic times, today it is quite inconceivable. But this interpretation is due to the fact that our society (and we have shown why) only accords social value to the specialised technician. It is no longer important to create polytechnicians, but to create polyconceptualists who are monotechnicians. No human technology in the set of human knowledge is isolated. It is essential that the fundamental concepts should be very widely diffused

196

and in a simple form which is understandable by all; and it is essential to show how they provide a general structure which offers a home for all the specialised branches of knowledge.

Some years ago people like Ivan Illich[45] were proposing a society without schools. They were accusing schools of not preparing people for life, of keeping young people apart from the world in a pre-alienated condition, and of separating them from reality and creativity. I would be the last person to oppose this, looking at all existing schools as they are. Illich proposes putting young people in contact with the world of objects at a very early stage: the world of the adult, the world of the tool and of technology. A recent tendency in French teaching, which no one could call revolutionary, has been inspired by this same motive; but this, in my view, shows that it is not revolutionary to put the child as soon as possible into contact with the world of occupational labour. "Earning thy bread by the sweat of thy brow" has never filled me with noble thoughts or moved me to deep emotion. Perspiration has never seemed to me a very enticing finality for man. Moreover, in order to make it a finality one needs a real movement of opinion, and even socialist realism, with its pictures of beautiful lads and lasses holding hands with sheaves of golden corn on their shoulders following a road that points towards the horizon (the socialist horizon, of course), has never made me tremble with aesthetic emotion. I am rather tempted to believe that "the lilies of the field toil not, neither do they spin".

All I have said in the preceding chapters shows, however, that I am making commendable efforts not to pass any value judgements on manual labour as against so-called intellectual labour. Certainly I think that there is nothing revolutionary about Ivan Illich's proposal to involve the child at an earlier stage in the production process so that he draws experience of the real world from it. I am sure that the English capitalists at the beginning of the industrial revolution, just like our capitalists today, would have very happily agreed with this opinion. But on the other hand, don't assume that I wish to defend the multiplication tables and problems about how long it takes to fill a bath, or even questions with multiple choice answers. They have the same end, though the method is probably less effective: to make the human being exclusively a commodity producer. If that is the aim, I am not interested in the methods used to attain it.

What I am interested in is a new reality in the consciousness

197

that we have of it: man is the only animal who knows how to process and create information by means of his associative brain. This is what must be protected from routine training and all conditioned reflexes; we must not only protect it, but also do all we can to help it develop. It is therefore urgent that before teaching children and adolescents to process mass, we teach them what information is, and how and why the nervous system which is common to the whole human species uses it. And we must teach them that this use of information can be a means of making objects, but that it is also and perhaps pre-eminently a means of knowing oneself and of determining one's place in the cosmic set, in relationships with others. The task of preparing children to face up to life and reality will not suffer as a result of this kind of education, but will on the contrary be helped by it. Why should the young person be left all on his own to rediscover (if indeed he succeeds in doing this at all) the hard road of abstraction which the human species has pursued from its earliest origins, at a time when this abstraction is already in a position to enable him to structure mass and energy on a daily basis? It is astonishing to read the numerous findings of commissions, subcommissions, conferences, meetings and symposia concerning educational reform. One sometimes reads the enlightening statement that a new, socialist society will give rise to a new kind of school. Then the question is immediately tackled as to what this school would be like: school structures are refashioned, and we are plunged into an analysis of what ought to be done and what ought to be undone. But the question of what is the finality and the content of education is hardly ever tackled. To be sure, there is always a corner reserved for "culture", that is to say art, and for sport. But science and technology are made to cohabit in such a way that it is pointless to use two different words, since the two things are never separated. Some contracted term such as "scitech" would be much more simple and would save time, and after all time is money. The process of cutting the loaf of human knowledge into neat slices goes on just the same. The sub-sets are reorganised simply by putting the commas in different places, within the same stunted frameworks. When, on exceptional occasions, the question "what is education for?" is asked, then either no reply is given, or the implication is that the purpose of education is to carry out occupational training plus a bit of artistic culture (rather like socialism being "soviets plus electricity"). In all, this will broadly suffice to fulfil the life of the human being, particularly if it is complemented by occupational retraining (sometimes known as "per-

198

manent education"). Biology, that is to say half the totality of human knowledge, occupies a sub-heading in the educational programmes, and is completely separated from the so-called human sciences — psychology, sociology, economics etc. — just as one learns to drive a car without knowing a single thing about the internal combustion engine. But at least a car usually functions without breaking down or misfiring, whereas one can hardly say as much for individuals and human societies. The basic principles of physics are reserved for career engineers, and will only be seen from the point of view of the production of machines and commodities. What use are the notions of entropy, structure, information, or their representation in mathematical symbols, to a man of letters? Like the rest of the world he is part of the stream of solar entropy, but he obviously doesn't have any use for the second law of thermodynamics, nor for the first law for that matter. People complain about the parcellisation of human labour, thinking of the unfortunate worker on the assembly line. But in reality there is not all that much difference between the assembly-line worker and the "intellectual" except for a difference in hierarchical status, and the reason for this we know. To paraphrase Bernard Shaw, knowing everything about nothing is not so very different from knowing nothing about everything, or even from knowing everything about everything like a well-stocked library. When are you going to be taught about the reason for living, and about open structures?

Some may say that in taking this line of approach there is a danger of "disengaging" or "demobilising" the masses. This argument is only true in so far as the masses find themselves engaged and mobilised on the basis of archaic schemas, of which those presented in this book constitute a critique. To argue in this way is to despise the masses, to believe them incapable of conceptual progress, to regard them as definitively conditioned by stereotyped language and worn-out slogans, and to regard them as irremediably motivated only by the needs of their stomachs and their appetite for consumption. To appeal only to the emotional susceptibility of the masses is to downgrade the individuals who make up these masses to the level of the species that came before us, and to "use" them to set up new forms of power and privilege.

Of course the masses have to be mobilised, but they have to be mobilised against every hierarchical structure of domination, against every structure that is closed, fixed, sclerotic, or analytic but non-synthetic; not only against those structures which currently exist, but also against those which might succeed them. And to mobilise and motivate the masses, it is far better to appeal to their reason than to their drives and cultural automatisms; at least, they need to be motivated in a reasonable way. Their basic drive must be to study the mechanisms by which their automatisms are established and the content of the latter. If humanity has one certainty today which the global "thought" of our great ancestors did not perceive, it is that action cannot continue to be guided by the instinctive hypothalamus or the limbic automatisms alone, but must control them, by means of the imagining cortex. Given this, then if revolution is necessary — why not?

12
War

There is a tendency in contemporary society, since ethology has taken over, to consider war as the result of an aggression which can already be perceived in the individual, and which can be encountered again at the level of organisation of social and national groups, and to consider that their common origin lies in the information-structure of the human species, in its genetic structure. I have already argued against this opinion (see page 49), and I have also studied it in greater depth in a recent book.[46] We know that aggression is not innate but constitutes a type of response to a stimulus in situations where this response cannot be expressed in an act of gratification, or when on the other hand this aggressive response has already supplied gratification by virtue of the domination which it affords. We also know that the finality of a living structure can only be to maintain this structure. Therefore, since every information-structure is a closed system, it is evident that in order to survive this information-structure enters into antagonistic competition with others for the appropriation of mass and energy. In the case of man and human social groups, this information-structure uses technical information in order to secure its domination. So we can define war as the result of a confrontation between two information-structures or closed systems; it takes place in order to establish the domination which is necessary for them to acquire their supplies of energy and material, which in turn are necessary to maintain these structures.

As the structure of all social groups, and particularly of nations, has till now always been a structure of domination, it follows logically that whatever the ostensible political, economic or energy-related causes, the aim of war has always been the maintenance of this structure of domination amongst the national groupings. We are talking about wars between nations, but this is no less true of so-called civil wars. The latter spring from the fact that within a national group two closed systems come into antagonism, an antagonism which often emerges because a third closed system comes on to the scene, acts as a detonator and, after the elimina-

201

tion of the institutionalised structure of domination, will take over the new phase of domination.

Aggression between nations in the form of war seems to possess mechanisms which are rather like those of aggression between two individuals. It springs, for example, from a given nation's inability to continue indulging in acts of (economic, etc.) gratification. Generally it is the dominant group in this nation which is unable to find gratification, but because this group controls the means of disseminating information, it can get the national set to believe that the private interest of this group is the national interest and so to win support for a "holy alliance" of all the people. So far, no horizontal opening between the functional classes of two nations on the point of conflict has ever been able effectively to oppose the outbreak of this conflict.

However, in certain cases the origin of the conflict would seem to be attributable to one dominant nation. Like a dominant individual, a dominant nation appears to have cast all its aggression aside. It is fully satisfied by its dominant position. But like the dominant individual it immediately has a tendency to extend its own information-structure, its way of life, and believes that the whole world should accept it, admire it and take part in it. Every socio-economic structure different from its own represents, for the dominant nation, a "barbarian" structure, and since its own structure has allowed it to assume domination, the others ought to imitate it and accept its "leadership". It also has a tendency to extend its economic ascendancy and to consider that all the goods of the world are owed to it, that they are its property. In exchange, it supplies friendship, protection and a few consumer trifles. Clearly, anyone who doesn't accept submission is a heretic, a "baddie"; he must be punished for threatening the peace of the world, which is its own responsibility. Any challenge, any attempt to get out of this economic or structural entanglement immediately invites warlike reprisals. These reprisals are always justifiable; the cause — broadly speaking, the cause of freedom — is just, for freedom obviously can only be achieved by acceptance of this domination. Finally, if violence has made it possible to establish a position of domination and hence of gratification, it will clearly tend to reproduce itself and become established as a method of gratification in itself. It follows that there can be no progress towards peace so long as the aim of social communications is (a) to maintain a human grouping's own hierarchical structure of domination by disseminating only the value judgements which aid the maintenance of this structure, and

(b) to reject the inclusion of this human grouping in an encompassing set which offers it no possibility of exercising domination or submission.

Once mass and energy circulate more or less freely between human groups, information must also circulate freely, that is both technical information and generalised information. Transnational sets as yet only consist of sub-sets of the human species, and information coming from sets outside them cannot circulate within them. This leads to the politics of blocs, which once again can lead to war. Universalism in the social structure of the human species would therefore seem to be indispensable.

Certainly it is easier to create national unity against an external, antagonistic system when the ranks of the hierarchical structure are staggered over many levels, as is the case in the industrialised countries. This is one of the reasons why it is so difficult to associate these closed national systems in a larger set. By the same token, in order to provoke a revolutionary crisis within such systems it is useful to spread the idea that there are only a few social classes, and to draw a veil over the multi-level nature of the occupational hierarchies: the social malaise can be utilised so that each individual, by virtue of his daily life, can situate himself in the camp either of the unsatisfied or of the satisfied, and thus it looks as if there are two closed and antagonistic systems. Conversely, in order to maintain a hierarchical system of domination it is advantageous to multiply the number of rungs on the hierarchical ladder, to multiply the number of closed systems, corporate groupings and sub-sets. This is simply making intuitive use of the old adage, "divide and rule".

To avoid war, therefore, it is necessary to open the social structures both vertically and horizontally. The history of the Jewish people is a very striking example of this. The diaspora was for them a horizontal opening extending across the peoples of the world. This horizontal opening was made, however, at a fairly specific occupational level (the arts, science, banking, commerce), a fact that to some extent reduced it and on the other hand facilitated the maintenance of its vertical closure. This information-structure of the Jewish people caused the national systems within which it settled to consider it an antagonistic system, an unknown which generated anxiety and every kind of evil and catastrophe. Much the same thing happened with the Protestants at one time. When the edict of Nantes was revoked, it was their horizontal opening which enabled many of them to leave their country and take refuge abroad.

As we sprawl on the cushions of our comfortable habits, should we then condemn the Palestinian commandos for their aggression and violence? They have no vertical opening with their "brother Arabs", nor do they have a horizontal opening with the peoples of the earth, because they lack sufficient technical knowledge and wealth. How can a human group find any gratification in dependence and submission if it has no "territory", no capital and no technical knowledge? The diaspora allowed the Jewish people to participate in the technical evolution of the peoples amongst whom they were dispersed. They were able to share in and benefit from the circulating information of these peoples. Vertical closure was the source of the successive attempts at genocide, of which they were the victims. But once they settled in the land of Israel, their horizontal opening enabled them immediately to benefit from accumulated international capital and in particular from the technical information in which they had participated. The Arab peoples on the other hand were never dispersed, remained vertically and horizontally closed, and generally stayed at a structural and technical stage of evolution belonging to the Middle Ages.

Every human grouping should audit its racial, linguistic, religious, ecological, economic, energy-related, cultural and suchlike "closures", so as to assess their "value" within the evolution of the modern world. It must determine whether its attachment to any of these values springs from the pursuit of domination or whether it comes from the refusal to submit. It must also determine whether the sacrifices to which people agree in order to preserve these values are disproportionate or not to the values themselves, values which at the level of the organisation of societies are merely the equivalent of the automatisms, prejudices and conditioning of its individuals. We need to ask whether many of these values should not be put in glass cases in museums, to be respectfully admired by the crowds, and whether other values permitting vertical and horizontal openings should not today be preferred and substituted for them. Unfortunately one also has to ask whether the laws of evolution themselves do not require every structure to pass through a stage in which it becomes temporarily encysted before opening more broadly on to the world. Would it have been possible to pass directly from unicellular creatures to man? Has evolution itself not been precisely such a succession of failures, errors and relative successes? How can a structure acquire con-

204

sciousness of itself before it exists? Was it possible to conceive of a world scheme of things without domination earlier on in history? Could the Pax Romana, like American peace and Soviet peace, have possibly represented anything but a prematurely closed structure of domination? Is it possible to hurry evolution?

On the other hand, if we believe as some people do that war is inevitable and even provides us with gains in the sphere of technological population control etc., then we are enclosing ourselves in a prehistoric structure, relying on past history to deduce the future for us. This is to remain within an Aristotelean system of linear determinism and infantile causality. It means failing to understand that the human brain, with its associative systems, adds an essential element to the brains of the preceding species. This essential element enables human beings to give form and to inform; it enables them to escape gradually from certain pressures of necessity and to discover the necessities of a higher level of organisation, instead of remaining imprisoned in those of the preceding levels of organisation. But conversely, if we try to resolve conflicts by organising what is external to man, then we deprive ourselves of a fundamental informational contribution and wilfully reproduce the errors of the past. It is curious to see how many of the mechanisms which come into play when an organism is in an "emergency" situation are similar to those used by a social group or a nation. I have dealt more fully with this idea elsewhere. We should simply note here that under such circumstances the whole dynamic equilibrium and habitual biological programme of an organism is *momentarily* sacrificed to flight and to struggle. The organs which make these two forms of behaviour possible receive a preferential supply of blood, proportionate to the metabolic effort or the work they must perform. But this redistribution of the stock of blood is only possible because certain organs not immediately useful in flight or struggle are deprived of their ordinary nourishment. If this organic reaction to aggression is not quickly effective, if it persists, it will cause damage to the organs which are not immediately indispensable but are secondarily indispensable to survival, such as the liver, the kidneys, the intestines and so on. From this a state of shock can result, leading sometimes to death. The transition from a peace economy to a war economy similarly encourages the activity of certain national functions and industries, at the expense of those which are not immediately necessary to the struggle. However, there is an important difference: in the organism, the organs which authorise struggle do not do so in order to obtain preferential nourishment, but in a social organism certain industries will seek to provoke war in order to develop at the expense of the national set. Clearly this is the result of the hierarchical structure and pursuit of domination which exist in the social organism but not in the individual organism.

The horizontal closure of these industries may disappear as a result of the vertical opening of national sub-sets into a planetary information-structure, but this is by no means certain. What happens to capital can equally happen to war industries: like the oil companies, they can become internationalised and form a vertically closed sub-set in a set which, at the level of national structures, is open. They will then tend to encourage the formation of other closed structures, and it is closed structures alone which are capable of conflict. The conflict may in fact spring from the closure of functional classes or structures which are already open horizontally, on a planetary scale, but which are insufficiently open vertically, in their relations with the other functional classes making up the structure and functioning of the planetary organisation. If the planetary organisation is to allow and perhaps even promote the regional individuality of human groupings in their respective ecological frameworks, it must prevent any human grouping from being self-sufficient or becoming isolated and closed in terms of energy, matter (raw materials) and information (information-structure and circulating, technological and generalised information).

13

Creativity

We have reached the conclusion that the fundamental characteristic of man is creative imagination:[47] not merely the imagination which creates commodities, but the imagination which creates new structures to enrich his knowledge of the world in which he is submerged. Let us therefore outline the mechanisms of creativity, in the context of the biology of behaviour.

Motivation

We make the supposition that the creator, like every living organism, is motivated. We also know that this motivation is essentially a pursuit of pleasure, that is to say of biological equilibrium. Finally we know that in the social situation, the satisfaction of the instinctive drives can only be achieved by means of domination. Domination in the social situation demands acceptance of the hierarchical system which has been institutionalised by the dominant of yesterday; it is then imposed upon the individuals of today. This perpetuation of the rules of domination displays a total lack of imagination on the part of the dominant: they are incapable of conceiving of other social systems, because the system which gives them honour also encloses them within its own socio-cultural automatisms.

A socio-cultural system which regards the commodity as king can only give birth to a hierarchical system centred upon commodity production; it thus castrates any creativity which does not lead to the creation of commodities. What motivation can the individual have in such a system except to ascend the hierarchical ladder? To do this, one must obey the rules of domination which guarantee the maintenance of the existing socio-economic structure. In such a system one is asked to reproduce, not to create new structures. One has to learn things and recite them, not to invent them (unless of course the invention is saleable and does not endanger the current socio-cultural system). Writers and philosophers are honoured *pro rata*, according to their support for the maintenance of the existing mental and social structures. What sells best is always that which creates the least disquiet, namely the expression of the commonplaces and prejudices of the epoch. In such a system much encouragement is given to teachers who transmit this cultural background without changing a word of it, i.e. to those teachers who adapt themselves to the examinations and competitions by which degrees of conformism are assessed. These tests reveal the power, authority and domination of those members of examining boards who know how to advance their own pupils and block those of their colleagues. This game of influence understandably proves quite attractive in university life; the "pupil" turned "master" later finds that he in his turn cannot do without it. Creativity in such a system is out of the question. The creator must find his motivation outside the hierarchies of the society in which he lives, for creation affirms a new structure, one which does not conform and which creates anxiety. This motivation cannot therefore be immediately accepted by the human grouping or society of the epoch in which it is expressed. It runs contrary to the hierarchies which establish their domination on the accepted basis: this conforms with the existing society, does not cause anxiety, and is useful for the maintenance of the structure of the group. But since creation is only possible outside the hierarchies, when it does burst through then the whole hierarchical world which constitutes the "unshakable" armature of human behaviour in society finds itself shaken. The fact that creativity can exist without the innumerable labels and "titles" with which the dominant reward themselves shows that these titles are a reward for mere conformism, not creativity. It is difficult for creativity to achieve satisfaction in hierarchies which reject it, and the hierarchies encourage conformism rather than creation.

The motivation of the creator can only be nourished by the existential anxiety which the hierarchical and paternalistic societies are compelled to conceal; however, this motivation must open on to a structure which is sufficiently general to be able to respond to this anxiety. If he allows himself to be diverted even momentarily towards hierarchical competition and the struggle for domination, then he runs the risk of being definitively lost for creation. He must waste no time in futile verbal battles and disregard both criticism and praise, unless they can help or enlighten him in the pursuit of his objective. On no account should he worry about his image in the eyes of others. The image will rarely be a favourable one; he is a nuisance, and nuisances are not rewarded with hierarchical rank-ings. If he is concerned about his image, he is in danger of getting sidetracked in the heat of battle, or of being depressed by the ignorance and contempt which he encounters.

Hierarchies become blurred beyond frontiers, for each of these frontiers demarcates a closed system. Each human grouping has its own hierarchy, a sociologically closed system from which the creative person must escape as quickly as possible. He must avoid accepting the finality which animates the individuals in this grouping, and which consists in trying to ascend its hierachy. Because he disseminates the results of his creative activity beyond the group to which he belongs, and because he is not at that point part of the hierarchy of the new grouping to which he is addressing himself, he will not make himself an object of fear: for he will not be competing for a position of domination. He will thus reach a larger set.

As soon as he can make himself heard beyond the frontiers, he is saved from being suffocated, gagged and castrated. Generally speaking he will still have to find some separate means of subsidising the pursuit undertaken by his group, for he can't expect any subsidies from the official quarters whose hierarchical rules he has failed to observe — either that, or he will have to arrange for his basic research to have some indirect bearing on commodity production, which essentially it has no connection with. In this case it is given the delightful name of "innovation", a delicacy which all the economic structures are fond of using. And it is a delicacy they don't often enjoy, for they fail to understand that it is precisely by not searching for the commodity that one finds it, that it is by searching for laws that one finds the applications of these laws.

The motivation of the creator has its origins in existential anxiety about death, the dark blackcloth of all human life, and it is therefore surprising to discover that he needs to disseminate the results of his creation. One would have thought that the construction of some comfortable imaginary world far removed from the harsh world of domination should in itself be reward enough. This is no doubt true for artistic creation. The life and work of Van Gogh shows this, to mention only one example. But even in the case of artistic creation the artist does not live in isolation, in the absolute; he lives in a society from which he demands food, shelter and clothing, in short the satisfaction of his basic needs (and not everyone has a Theo to help). The true creator submits to this economic alienation as little as possible. But because there is no absolute criterion of beauty in art, the arbiter of success is a complex mixture of motivations: those of the artist, those of the dealer who makes a profit from his work, those of the person who looks at it or buys it. This success is social success: that is, a hierarchical or economic success.

Unfortunately, the difficulty of assessing criteria of beauty in art leads either to the rejection of any originality which might disturb the automatisms or to the admiring acceptance of some kind of eccentricity. Eccentricity is not necessarily artistic. A person who, in the world of the market place, has no great gifts for abstraction in occupational information, finds it difficult to establish his domination and therefore to achieve gratification. In art, he finds an exceptionally good terrain on which to express his aggression, in the form of an eccentricity which cannot be grasped by the systems of reference. Many of those who contemplate an uncomprehended or (more precisely) incomprehensible work will not want to appear not to understand it; and besides, the appearance of understanding it places them outside the common rut. Outrage, even without art (which normally is something impossible to acquaint oneself with on the spot), will always find a snob public to appreciate it. Only time allows a general consensus to insert a certain eccentricity into the domain of art. This eccentricity is difficult to assess *a priori* in a coherent way. Art is the result of taking soundings *a posteriori*. Art is always eccentric at the moment of creation, but the converse is not true: eccentricity is not necessarily artistic. A hold-up or a violent assault on accepted mores does not necessarily make a person an artist.

It is conceivable, at least, that artistic creation can be self-sufficient, that it has no need of a public. Public recognition makes it suspect. If the creator presses his work upon the public, it casts suspicion on his motivation: his flight into the imaginary has no need of that public which is sought by anyone unconsciously attempting to find a place in the hierarchies.

The same may be said about scientific creation. But here the common excuse for the attempt to gain public recognition is doubt about the value of creativity's approach to the real. To be sure, experimental control by others is a check on scientific fictions, and such control must be obtained by disseminating results. But the motivation to be the first to discover an unknown aspect of the world springs from congenital narcissism, from the need to be loved and admired by which every creator is led on. It is only when he sticks to what he is doing in the absence of such rewards that one's suspicion diminishes.

It is only in the absence of such rewards, in the absence of immediate public recognition for his work that its originality becomes probable and his motivations are likely to be basic. But then he also shuts himself up in the closed system of his imagination; and even though he may consent to experimental control, one can still see that the distance separating the aggressive personality, the rebel, the drug addict and the psychotic from the artistic or scientific creator is quite small.

Memorised experience

We have seen that the sociocultural automatisms are based essentially upon memorised experience. The consequences of these automatisms can only be unfavourable. An automatic act, being unconscious, makes action more rapid and effective than if it had to be rediscovered or thought out anew on each occasion. A society which is entirely orientated towards commodity production seeks to encourage these automatisms at all levels of the occupational hierarchies, and to establish them in a stable manner. In our daily lives the automatisms populate our behaviour. We walk, we breathe, we drive the car, we behave in society in an almost entirely automatic manner. In addition, our actions are fashioned by the automatism of language, the expression of the "culture" of the moment. When we proceed from the deed to the spoken word and examine how these sociocultural automatisms are established from birth in any given society, we are forced to admit that almost all our judgements are simply the expression of automatisms of thought, which only have a value by virtue of the conformism necessary for the survival of an individual in a given society.

This leads to several consequences. The first is that while the conditioned, conforming subject may find his biological equilibrium, pleasure and reward in the hierarchical system and so-called moral values imposed upon him by the social structure to which he belongs, there are other cases where these automatisms oppose the instinctive drives and create synaptic conflicts, since the subject cannot find satisfaction by effective action upon the environment. One can create this kind of experimental psychosis in animals. The same thing probably happens in human beings. Here the resolution of the conflict can only be found in the aggressive outburst, neurosis or psychosis (depending upon how far back the conflict goes), escape into drugs (of which alcohol is the most widely used) or make-believe, the creation of another world which offers consolation and rewards. Artistic and scientific creation correspond

211

to this last solution. But the motive we encounter here is not only the pursuit of pleasure, it is also the knowledge that it is impossible to achieve pleasure in a specific system of hierarchies and domination; it is not only an escape from the pain of being subject to unacceptable laws, but an escape towards a new structure which itself is rejected by these laws.

But the new-born child can create nothing, in the sense that we understand "creation" in the scientific disciplines: it has memorised nothing, its experience is nil, and its associative activity, having nothing to associate, cannot yet lead to any activity which is creative of original structures. Clearly, this creative activity will increase as it is exercised upon a greater amount of memorised material. This means that the individual must pass the best part of his time gathering information. But the information-gathering demands the observance of two fundamental rules.

The first rule, as we now understand, is: not to be motivated by the pursuit of hierarchical reward. If in fact this is the motivation, it can only conform with the existing conceptual frameworks, and this obliges it to be monodisciplinary rather than interdisciplinary. In fact the only person who is rewarded in a technological society is the specialist: he is the only one to reap the reward of university and academic honours, etc. We shall examine the advantages and disadvantages of interdisciplinarity from the social point of view a little further on.

The second rule is to introduce newly acquired knowledge into an open nervous structure, by means of a methodology which I outlined when I was speaking about "generalised information". If information penetrates a closed structure it may well be rejected as non-significant or, even if it is accepted, it may well not provide any effective material for the imagination to work upon, it may be throttled by the iron yoke of conceptual automatisms. "A fact is nothing by itself. It is of value only by virtue of the idea attached to it or by the proof it supplies" (Claude Bernard). "The idea" in this case represents the structure, and this structure only permits a certain progress if it is an open structure.

Now we have seen that the opening is a result of information coming from outside the system at another level of organisation. Thus an "idea" is fundamental to the extent that it maintains its coherence with experimental facts gathered at other levels of organisation, i.e. to the extent that it constitutes a model which can be used from an interdisciplinary standpoint. We may go further. Experience proves that an idea or an imaginary model which is

212

applicable at many levels of organisation and allows for the variations induced by the workings of different servomechanisms, subsequently has a much better chance of being proved experimentally than a model which can only be used at one level of organisation. In the course of research, the use of this criterion of effectiveness marks a saving of time and effort.

Unfortunately at the present time there is a plethora of pseudo-syntheses which are stuck together with scissors and paste. These are the mere interdisciplinary juxtapositions which sometimes get accepted — this is particularly so in the domain of the so-called human sciences, where they can be confused with the creative syntheses of new sets. In such compilations the whole represents the sum of its parts. All new structure is excluded from it. This kind of general survey can sometimes be useful, but one could say that the fate of the human species does not hang upon it.

At all events, the information-gathering is a burdensome task, a long effort which must be constantly renewed. Genius is never spontaneous. Only a powerful motivation is capable of inspiring this effort, and only an open structure is capable of turning it to advantage. It is also necessary for the finality to be somewhat larger than the size of a buttonhole.

Memorised knowledge is the indispensable material for creation. We sometimes call it a "craft". It is the craft of the pianist, for example, endlessly repeating the same piece so that the movement of his fingers becomes unconscious, a conditioned reflex, and so that he can then direct his attention to the tone, with which he adds his own emotion and imagination to the work he is interpreting. It is the craft of the painter, the sculptor, the architect, the actor, the biologist, the mathematician, and the physicist. It is the craft acquired in "training" which enables us, before we embark upon unknown paths, to know precisely this — that they are unknown — because the paths previously opened up have already been explored. I repeat: training is indispensable, but it is also dangerous, because it gives a sense of certitude, of knowing everything, and because it shapes the nervous system so rigidly that it is often very hard to plot a new network of synaptic relationships with such a firmly fixed network as one's starting-point. A person who has only one craft has little chance of becoming a creator. But if there is no craft at all, there is certainly no creator.

The complexity of methods and the large amount of technical knowledge which it is necessary to acquire within a scientific or artistic discipline in order to master its practice, are enough often to fill a lifetime. The specialised techniques which force the individual to measure himself against the constraining rigidity of facts, the strict laws of energy and matter, are a school in humility for the creative person and a safety device to prevent him from being swept away by his imaginative activity alone, without control by the surroundings or any reference to the facts. But equally they should not constitute the grid over a conceptual and praxic cage from which he cannot escape. A craft is thus also necessary in order to understand that the laws which govern living processes bend before the imagination no more than those which govern inanimate matter. But it is questionable whether this experience is necessary in all the disciplines which are going to occupy the researcher's memory, for one person's lifetime is not long enough to acquire it. The conductor of an orchestra regards himself in a poor light if he has not first been an instrumentalist, and a composer usually knows how to play an instrument, but it is not essential that he should know how to play all the instruments in the orchestra. Creativity has perhaps a much greater need of polyconceptual monotechnicians than of true polytechnicians. Now we have seen that the most useful way of handling polyconceptuality is to possess an open mental structure, which is capable of organising knowledge without at the same time enclosing it. Once again we encounter the distinction between thermodynamics and information. The technician of a discipline uses his information to effect thermodynamic action, to bring a technique into play; but the polyconceptualist can deal only with information, once his experience of thermodynamics within one discipline has taught him to link the deed to the word and to harmonise them.

If we consider the specialist as a set, creativity will most frequently spring not from the establishment of new relations between the elements of this set but from the meeting or intersection of this disciplinary set with another. The new conceptual models thus created will make possible the pursuit and the discovery of new elements and new structures which will enrich knowledge and make action more effective.

The need for synthesis after a historical period of analysis has made the interdisciplinary team fashionable. Anyone who has had the responsibility for forming such groups and has taken part in them knows that the job is not easy. The first reason for this is a semantic one. The languages of each discipline serve as a means to convey specialised information, and are sometimes different in the words they use for the same object, since this object is seen each time in a particular relationship. For example, the language used to describe one and the same molecule by the physicist, the chemist, the enzymologist, the biochemist, the biologist, the pharmacologist, the physiologist or the clinician, does not lead to a fruitful exchange of information on this topic, for the language is too different from one discipline to another, unless each of them makes a painstaking effort to acquire an interdisciplinary language. This interdisciplinary language needs to be linked to concepts, rather than to the techniques used to understand this molecule in one particular discipline.

The second reason stems from the fact that monodisciplinarity is a sort of "territory" in the ethological sense of the term, or more simply a private game reserve. Anyone who trespasses, with man as with the animal species, releases a reaction of aggression against himself. Moreover the trespasser, the stranger to the discipline, is bound to be ignorant. The vision which the specialist has of an object fills his field of consciousness to such an extent that he finds it hard to imagine that this personal vision of this object cannot constitute the entire and sole reality to be drawn from it. So the distress which springs from the presence of another, of something different and unknown, can unfortunately only be resolved by negative rejection or by an attempt at paternalistic domination.

Once all the obstacles have been surmounted, an interdisciplinary team can certainly be an effective instrument of discovery. There is, however, a last obstacle still to be surmounted. It consists in the fact that such a team generally forms part of a very much larger socio-economic set. The desire for hierarchical promotion, both socially and on the income ladder, often leads to a change in the basic motivation: this is no longer discovery but the conformism required for suitable appreciation according to the habits, prejudices and value judgements of the moment or (worse still) of a superior hierarchy. The existence of hierarchies in general is, certainly in some countries, the disease which most effectively kills off creativity, not only from interdisciplinary teams but also from the researchers themselves. A researcher is always paid by someone or other.

I accept that creativity is clearly the finality of interdisciplinarity, and that it springs from the original association of elements taken from different disciplines. It is nevertheless difficult to know whether these associations should be achieved by bringing together different individuals from several different disciplines, despite the inherent difficulties resulting from the dynamic of human groups, or whether these associations can be realised in one and the same individual, whom we have called a "polyconceptualist" in order to contrast him with the "polytechnician".

The study of creativity here touches on sociology, for the most desirable solution is a team of polyconceptualist monotechnicians. They must be monotechnicians because it is certainly necessary at some moment or other to "inform" matter, and because the ability to give form to matter is, as we have said, difficult for one person to acquire in several disciplines at the same time. They must be polyconceptualists because this is the simplest way of getting rid of the interdisciplinary morgue, and of allowing the "languages" to facilitate a profitable exchange of information among monotechnicians. This could lead to the creation of new structures in which everyone can participate at the conceptual level, since each person will find himself protected at the same time by the technical ability of all the others.

This kind of team would constitute the model, perhaps a utopian one, for a harmonious and effective human society based on the necessary distinction between specialised information — which could also be called occupational or technical information — and generalised information, which concerns the structures as a whole and their functional dynamism. Is there any other way of achieving tolerance?

The imagination

I said that a newly born child is incapable of creation: it has no memorised experience which is capable of supplying it with the material necessary for the expression of its associative faculties. On the other hand, it needs to be emphasised that memorised experience cannot be present at every moment in the researcher's field of consciousness either; the greater part of our memory represents material which has departed from this field of consciousness for a certain period of time, and so it cannot always be easily recalled. However, daily experience in the activity of research demonstrates that this unconscious experience (provided, I repeat, that it is not too narrowly imprisoned by the acquired automatisms) is apparently capable of acting as a substrate to the associative processes. The imaginative structures to which it is then able to give birth, coming to the level of consciousness in a secondary way, appear to be a free gift from some benign goddess: intuition. In reality, intuition demands the long-term effort and painstaking work of information-gathering. But this information-gathering perhaps does not necessarily mean that the memorised experience has to be maintained in the field of logical discourse. If this hypothesis is admitted, as many experimental facts in the domain of neurophysiology would tend to suggest it should, one should even more determinedly distrust conscious logic, which is capable in certain cases of stifling unconscious logic. Unconscious logic is not in fact the logic of discourse but rather that of the complex biochemistry which governs the activity of our cerebral neurons.

If we accept this point of view, then we have one more argument in favour of interdisciplinarity and open structures as the guides to the individual's information-gathering.

Creativity: why?

It may seem strange to devote a chapter to creativity in a book which aims to be as general as this one. Isn't this a problem which concerns only a small number of people and consequently lacks major interest for the species? In fact, all the progress made by this species since the beginning of human history has been the result of such creativity. It is true, however, that so far it has remained the private domain of a few privileged beings, generally by virtue of their birth and by the happy coincidence of an immediate environmental niche which favours creativity. But the creator's rarity is regretted, and this itself lends support to the idea that the life of every human being could be creative if the social set supplied him with an adequate framework for nurturing his imaginative faculties. I have repeatedly demonstrated that if there were a nonhierarchical motivation, if there were no intransigent automatisms whose goal was quite plainly to encourage commodity production and productivity, and if every person had free time available to him outside his occupational activity (whatever that occupational activity might be), then it is probable that a very large number of individuals would become creators. All the knowledge which I have dealt with above about the mechanism of motivations, the establishment of hierarchies and the meaning of what I have called generalised information, can be used for the development of human creativity. The attribute of creation, of creating information by means of the imagination with memorised experience as the starting point, is possessed from the day of his birth by every human being who is not mentally handicapped. If he loses this attribute it is his environment which is responsible, whether this environment be the socio-cultural niche of a bourgeois family, of "intellectuals", or of a large set of workers. In the first of these instances promotion in the hierarchy will certainly be much easier than in the others, but this will not help his creativity. So for a long time "equal opportunity", a highly logical demand, will be simply the equal opportunity to become a bourgeois.

We should feel deep pity and sympathy for all those whose imagination has been castrated, and who vent their spleen and bile against the marginal character who doesn't take the straight path. If they curse him or pretend to ignore him, it is because they are vaguely conscious of their impotence, because they sense in some obscure way that their lot is not the best, and they are stuck with it. These are the true victims of social destiny, and they need pity.

14

The finality

I have repeatedly said that the human being of tomorrow will
have to change his finality. But we also know that both the individual and the social grouping only have one finality: the maintenance of their structure. Consequently when I speak of changing
the finality of the human being of tomorrow, I seem to be throw-
ing myself into a contradiction. Actually I have been taking a short
cut which made the exposition of the argument easier. What I
ought to have said is not that he will have to change his finality in
order to survive (since his finality is precisely survival), but that he
must change the means which he has so far employed to secure this
finality. The processing of information, which over the ages has
enabled humanity to survive as a species, has so far been used
either as an individual means or as a group means of establishing
domination. It is the diversion of this extraordinarily powerful
means of acting upon the surroundings into an individualist or
group project that is now leading man to ruin, and to the malaise
whose origin is so remote that he can no longer clearly perceive
the determinism behind it.

According to some the future is already lost, since aggression
and the search for domination are inherent in the genetic inheritance
which has been transmitted to the human being by preceding species.
I have already expressed my views about this so-called "congenital aggression" (see page 46). But in fact the human being is also
endowed with an imaginative consciousness. He is thus certainly the
one species to have understood today the danger which is posed to
the species by the pressure of necessity, a pressure to which innumerable species have been subject (together with his own) since
the beginning of time. Once he is confronted by the imminent fear
of disaster, once he has unsuccessfully tried all the petty means that
technology can provide to delay the disappearance of the hierarchical systems of domination, which are at the source of the destruction of the biosphere and of the inadequacy of the social systems,
perhaps then he will change the means he has so far employed to
survive. We ought rather to talk about "the means he has so far

employed to maintain his social structures" and the hierarchical structure of his societies. The survival of the group ceased to be linked to domination once the population began to grow and the life of all the human beings on the planet could be affected by the action of a single human group. If humanity wishes to survive as a species, it must behave as such and not as a partisan of one particular group.

The pursuit of well-being through consumption is a powerful motivation, and can be achieved if the individual belongs to a human group. This group, by means of its domination over other groups, can satisfy such a motivation even if the hierarchical system within the group involves a certain social malaise. In a planetary society, on the other hand, if well-being through consumption still exists (as doubtless it will), it can only be an indirect by-product of behaviour, not its essential end.

What motivation can the human being of tomorrow discover, if he wishes to ensure the survival of the species? I have proposed elsewhere[48] that his aggression be diverted away from the human environment and towards the inanimate environment. The organic reaction to aggression (ROPA — *réaction organique postagressive*) makes flight or struggle possible against a fierce beast and then against an enemy invading the territory, but it is no longer any use when it is harnessed against the boss, the foreman or the neighbour across the landing; these are people from whom it is no longer possible to fly or whom it is no longer possible to put to flight. In the same way, the aggression which expresses this in behavioural terms is generally speaking useless in the close-knit sociological network which imprisons the citizen of today. The motivation will always be the pursuit of pleasure. But man must learn to satisfy this motivation no longer in the mere acquisition of occupational knowledge, nor in some form of social advancement based on the rules of an occupational hierarchy of domination, but in creativity, in the possession of political power by the functional classes, and in the acquisition of generalised information. He must be politically motivated. Politics must become the fundamental activity of human beings.

But we must be clear about what we mean by "politics". I have no intention of reducing it to the meaning with which it is widely used today, as in the phrase "I don't get involved in politics, you know". It is not a question of subscribing or not subscribing to the ideas held by a party, still less is it a question of accepting or criticising the actions of politicians. On the contrary, the point is to understand the general bases of human behaviour in the social situation, the causes of the present structure of human societies, the economic and cultural relations between these societies, and their mechanisms. To take the argument to the point of paradox, I would be tempted to say that biology and politics ought to be virtually synonymous. The point is to make politics a science which is not just a matter of words, but one whose conceptual models would be sufficiently open to all the contemporary scientific disciplines, so that when experimentation takes place it does not lead to a dead end, which is what has always happened up to now.

The human being of tomorrow must be motivated to understand that he can only find security, gratification and pleasure by being concerned about others, or more precisely about the relationships of human beings to one another, all human beings whoever they are. I rather believe we are entering an era when it will no longer be possible to be happy on one's own or with just a few other people. We are entering an era when all the old "values" which encourage hierarchical domination must collapse. Moral rules, laws, labour, property, all the regulations which smack of the barracks or the concentration camp are the consequence of humanity's lack of consciousness, which has led to imperfect socio-economic structures in which domination needs police, armies and states to maintain itself. As long as all the various forms of coercion persist, they will be proof of the imperfection of the social system which has need of such things in order to keep going. As long as human beings seek to impose their truth upon other human beings, we shall have inquisitions, stalinist show trials, morality, policemen, torture and the debasement of the human brain by the darkest prejudices of our unconscious and pre-human motivations. The future which I am suggesting seems too beautiful to be realisable. But logical reflection enables one to find solid arguments for stating that it is possible. Once economic evolution (that is to say man's technology, the fruit of his imagination and his accumulated experience) makes it possible for him to use materials and energy effectively in order to satisfy the basic needs of all human beings, then if distribution is properly carried out, all his other needs are socio-cultural and acquired by training. The idea of ownership of objects and beings, the idea of hierarchical domination determined by the degree of abstraction in occupational information, the idea that the work of producing consumer goods is the first necessity, the idea of social advancement, of equal opportunity to consume: all these ideas are simply training, going right back to the primary source of each

221

individual's socio-cultural automatisms, the family. For these solidly established ideas no longer to make sense, it is simply enough to learn something else, from earliest infancy. It would be sufficient for us to replace these automatisms with generalised information, the methods and objects of which I have sketched out above; if we did this, everything would change. You may ask whether I am sure what this would lead to. My answer would be no. This may seem paradoxical. But to believe that the disappearance of these old automatisms and their replacement in the human nervous systems by an open structure, a real culture (a system capable of organising all the sets and sub-sets, with all the elements entrusted to it in such a way that they could never be imprisoned in value judgements and socio-cultural automatisms) can make man a less developed creature than he is at present: that would be a much greater paradox.

So why not try? Simply because the hierarchical structures of domination everywhere in the world will prevent this state of affairs from being realised. Then they would lose their domination.

So it remains only for us to wait for the pressure of necessity to reach terrible proportions, for the determinism of evolution to take charge of the situation or not to take charge. . . .

The pressure of necessity has made evolution possible, by favouring first of all the strongest, then the most technically developed. Perhaps before long it will favour the most conscious.

But will intense fear by itself save the human species, forcing it to accomplish this structural leap?

If anything lasts in the human being, it is the pursuit of pleasure, since this — for the human being as for every living creature — is the means of keeping alive. Perhaps I have not sufficiently emphasised the idea that the avoidance of pain is the equivalent of gratification and pleasure. They are, however, two different kinds of behaviour, for to avoid stimulation of the periventricular system (PVS) is not the same thing as to seek stimulation of the medial forebrain bundle (MFB). When flight is seen to be impossible, avoiding pain can lead to revolutionary aggression, whereas seeking to maintain pleasure and to gratify it by domination can lead to conservative aggression. It all depends upon the point of departure, and we can see here how incomplete the behaviourist approach towards aggression can be. But in either case, man has so far found only one solution: to obtain domination, to seek power. All politico-economic ideologies have always made him believe that he is nothing but a "producer" of objects; it is within the framework of the process of production and of its hierarchies that he has sought domination. This means that the more he seeks to dominate, the more he produces. But it also means that this production is distributed in a hierarchical way, since the pursuit of hierarchical domination constitutes the basic motivation. The vicious circle is closed: it can only lead to economic expansion, to mitigated dissatisfaction in the industrialised countries, to mute and fatalistic suffering in the underdeveloped countries, to increasing pollution and to the death of the species.

It is necessary to get rid as quickly as possible of the idea that man "is" a workforce. He is a structure who processes information, and who also turns out to be a new source of information. It is quite permissible that a part of this information should serve the transformation of matter and energy and should lead to the creation of objects which above all have use value, before exchange value (which serves to maintain domination). But the fundamental error fostered by contemporary socio-economic ideologies is that this information, which the imaginative brain of a human being generates, should serve exclusively to produce objects and commodities. It is high time for man to realise that the main object of this information ought to be the creation of social information-structures which are no longer centred on the process of material production. Since mass and energy have lost most of their secrets, it is the secrets of the biologico-social information-structure that ought to be our most pressing preoccupation. After spending centuries scientifically (that is to say experimentally) studying inanimate matter, it is time to start studying, teaching, generalising and disseminating the structural laws of living matter, and to extend these studies to human sets.

Perhaps new structural laws will thus be found which apply to social constructions and not only to economic ones, although one should remember that it is impossible to dissociate information from mass and energy. Once a means has been found to prevent the social stimulation of the PVS of large numbers of people, then perhaps it will be possible to discover a means of stimulating the MFB in some other way than by hierarchical domination. In my view, this is where it must begin.

However the human being will continue to be born, to live, to make love and to die. The ever-present anxiety of death will still stimulate his PVS. But instead of being hidden as it is at present in order to protect the process of commodity production, this moti-

223

vation for action may well at last encourage a thirst for knowledge. If the biological and human sciences have demonstrated their effectiveness by basing themselves on the sciences of matter and structural laws, it is possible that the human being of tomorrow will abandon the myths that give him a feeling of security, that he will abandon his fatalistic submission to an incomprehensible determinism and his flight into drugs. In the fully conscious human being drugs might perhaps even encourage the imaginative, without leading to dependence.

So what may save the human species is perhaps not so much the fact that "man is a tool-making animal", but rather the fact that he is an animal — the only animal, of course — who knows that he has to die.

15

Epilogue

The sun is going out and the planet is dying. The sun, with its brightness and warmth — the sun of fire and light, spring and flowers, summer and fruits — finally uses up its energy; the final explosions burst forth in its once incandescent atomic hearth, and the fires of life and joy breathe out their last. On the planet there is a strange being, composed of many millions of cells. None of these cells understands the language, because it exists in a space which is too complex for them to participate in it (by the same token, the cells of our lives do not participate in our own human discussions). Time and space no longer means the same thing for the strange being as they do for the individuals who make up its elements. And yet they know about all these events concerning their common fate: information circulates among them without noise interfering, it is broadcast in three dimensions at the speed of light.

Because our time and space have no meaning for the strange being, it does not need arms and legs in order to act in this different space and in the duration of this other time. It has no need to act at all: it communicates. It has taken more than a million years (of our time) to be born to consciousness, a consciousness which human beings know nothing of. Its internal medium is the biosphere of the planet, but a biosphere which has already been profoundly transformed, in which all the individual cells are swimming. Its environment is the galaxy. The question is, will the galaxy be able to supply it with the energy it needs in order to maintain its cells and the stability of its internal medium? Will energy reach the dead sun from spaces which we cannot at the moment even conceive of, while matter spins round as a great planetary body in this information-structure, the information-structure of humanity? And how long will this take in our human, unidirectional and irreversible time, the time of our clocks? The cells continue to reproduce, to live and die. But the strange being lives in a different time-span.

On other planets, other beings like it have slowly built up, and the set of these beings is like a vast community. Many millions of

225

them have developed over many millions of years, and have died as it too will die, a passing cell in the living body of the galaxy. And this living body of the galaxy, which has transformed and informed inanimate matter into living matter, is in turn only a single cell which is born, grows and dies, a cell in the cosmic body which is all eternity and all consciousness; but this consciousness is inconceivably distant, total and totally different, it is a consciousness for which time and space no longer exist as others and no longer exist at all, because it is time and space itself. It does not have to be displaced for a certain period of time in order to cover a certain distance: it is what it is. Everything is finally enclosed within it. No doubt it is curved, instantaneously expanding, and cyclic; but it is also eternal, a duality of matter and anti-matter. And what does it serve in turn?

Notes

1. Henri Laborit, *Les comportements: biologie, physiologie, pharmacologie*, Paris 1973. See also Laborit, *Réaction organique à l'agression et choc*, Paris 1952.
2. Laborit, *Société informationnelle: idées pour l'autogestion*, Paris 1973.
3. I produced this hypothesis in 1961, before von Foerster had demonstrated the possibility of creating order on the basis of noise (see Laborit, *Physiologie humaine, cellulaire et organique*, Paris 1961).
4. C. S. Pittendrigh, in A. Roe and G. G. Simpson, *Behaviour and Evolution*, New Haven 1958, p. 394.
5. I developed these ideas on the cellular organisation of the nervous system more fully in *L'agressivité détournée*, Paris 1970. This is a non-specialist work. For those who wish for specialised information, I would advise *Les comportements* etc., in which there are more than two thousand bibliographical references constituting a very ample source of inter-disciplinary knowledge.
6. Laborit, *Les régulations métaboliques*, Paris 1965.
7. Laborit, *Biologie et structure*, Paris 1968.
8. P. D. McLean, "Psychosomatic disease and the 'visceral brain' ", in *Psych. Med. II*, pp. 338-53 (1949).
9. B. Milner, S. Corking and H. I. Teuber, "Further analysis of the hippocampal amnesic syndrome", in *Neuropsychologia* 6, pp. 215-34 (1968).
10. Laborit, *Les comportements* etc.
11. H. Hyden and P. Lange, "Protein synthesis in the hippocampal pyramidal cells of rats during a behavioural test", in *Science* no. 159 (1968).
12. B. F. Skinner, *Behaviour of Organisms*, New York and London 1938.
13. Laborit, B. Calvino and N. Valette, "Action du chloramphénicol, du bromure d'éthidium et da la N-formyl-methionine sur l'acquisition et la rétention d'une réflexe conditionné d'évitement chez le rat", in *Agressologie* vol. 14, no. 1, Paris 1973.
14. J. Olds and P. Milner, "Positive reinforcement produced by electrical stimulation of septal area and other regions of rat brain", in *The Journal of Comparative and Physiological Psychology*, 47: pp. 419-27 (1954).

15. D. L. Margules and L. Stein, "Cholinergic synapses of a periventricular punishment system in the medial hypothalamus", in *The American Journal of Physiology* vol. 217, no. 2 (1969).
16. W. B. Cannon, *The Wisdom of the Body,* New York 1932.
17. Claude Bernard, *Leçons sur les phénomènes de la vie communs aux animaux et aux végétaux,* Paris 1878.
18. Laborit, *Les comportements* etc.
19. J. Trueta, "La circulation rénale et sa pathologie", in *Mém. Acad. Chir.* vol. 74, no. 31-32, Paris 1948.
20. H. Selye, "A syndrome produced by diverse noxious agents", in *Nature* vol. 138, no. 32, London 1936.
21. H. Selye, "Stress: the physiology and pathology of exposure to stress", in *Acta Inc. Med. Publish.,* Montreal 1950.
22. Laborit, *Réaction organique* etc.
23. J. M. R. Delgado, "Aggression and defense under cerebral radio control", in *Proceedings of the Fifth Conference on Brain Function, 14-16 November 1965 (Forum of Medical Science,* Los Angeles 1967).
24. R. Plotnik, D. Mir and J. M. R. Delgado, "Aggression, noxiousness and brain stimulation in unrestrained rhesus monkeys", in B. E. Eleftheriou and J. P. Scott (eds.), *The Physiology of Aggression and Defeat,* London and New York 1971.
25. A. S. Welch and B. L. Welch, "Isolation, reactivity and aggression: evidence for an insolvement of brain catecholamines and serotinin", in *The Physiology of Aggression and Defeat* (op. cit.).
26. Laborit, *L'homme et la ville,* Paris 1971.
27. Laborit, *Résistance et soumission en physio-biologie, l'hibernation artificielle,* Paris 1954.
28. S. Boyden, "The impact of civilisation on human biology", in *Aust. J. Exp. Biol. Med. Sci.,* 47, 3, pp. 299-304 (1969).
29. G. Schaller, *The Mountain Gorilla,* Chicago 1963.
30. E. M. Thomas, *The Harmless People,* London 1959.
31. K. Jhamandas, J. W. Phillis and C. Pinsky, "Effects of narcotic analgesics and antagonists on the *in vivo* release of acetylcholine from the cerebral cortex of the cat", in *The British Journal of Pharmacology,* London 1971.
32. G. Illiano, G. P. E. Tell, M. I. Siegel and P. Cuatrecasas, "Guanosine 3', 5'-cyclic monophosphate and the action of insulin and acetylcholine", in *Proc. Nat. Acad. Sci. USA,* 70, 8, pp. 2443-7 (1973).
33. J. Piaget, *The Psychology of Intelligence,* New York 1950.
34. Lewis Mumford, *The City in History,* London 1961.
35. M. Cabanac, "The physiological role of pleasure", in *Science* 173, pp. 1103-7 (1971).

36. M. Cabanac, *op. cit.*
37. Delgado, *op. cit.*
38. Laborit, *Société informationelle* etc.
39. Laborit, *L'homme imaginant,* Paris 1971.
40. Laborit, *L'agressivité détournée* etc.
41. Gérard Mendel, *Sociopsychanalyse II: la plus-value de pouvoir,* Paris 1972.
42. *Ibid.*
43. Alvin Toffler, *Future Shock,* New York 1970.
44. Laborit, *Du soleil à l'homme,* Paris 1963.
45. Ivan Illich, *Deschooling Society,* London 1971.
46. Laborit, *Les comportements* etc.
47. Laborit, *L'homme imaginant* etc.
48. Laborit, *Réaction organique* etc.

Glossary

ADENOSINE TRIPHOSPHATE (ATP)

The combination of a puric base, adenine, with a sugar with 5 carbon atoms (pentose) gives a nucleoside, adenosine. The addition of a phosphoric acid to this nucleoside gives a nucleotide, adenylic acid or adenosine monophosphate. Two phosphoric acids give adenosine diphosphate. Adenosine triphosphate (or ATP) contains three molecules of phosphoric acid. Release of the last two molecules of phosphoric acid leads to an energy release of about 8,000 cal. Their bond with the molecule is called energy-rich and symbolised by the sign $\sim$.
It is written in this way:

$$\text{adenosine} - \underset{O}{\overset{OH}{P}} - O \sim \underset{O}{\overset{OH}{P}} \sim O \sim \underset{O}{\overset{OH}{P}} = ATP$$

This energy, which can be released easily, will be used by the cell to maintain its structures as well as for its synthetic or functional activity. One of the main functions of the chemical factory which a cell represents is the synthesis from food substrates of molecules of ATP. This synthesis is carried out with a low yield (2 molecules of ATP for one glucose molecule) by glycolysis, and with a high yield by mitochondrial oxidative phosphorylations (38 ATP for one molecule of glucose).

AMYGDALA OF CEREBELLUM, AMYGDALIAN NUCLEUS, OR ARCHISTRIATUM

Together with the hippocampus it forms part of the old brain, and sends fibres to the nuclei of the septum. The whole forms a vital relay centre of association between the rhinencephalon, the visceral cortex, the thalamus and the hypothalamus, which join together under the name of "circuit of Papez".

CATECHOLAMINES

A term designating the hormones of the adreno-sympathetic system and their urinary metabolites. This term recalls the diphenolic structure of adrenalin and the presence of an amine function on the lateral chain. There are three principal biological catecholamines. *Dopamine* is found especially in the brain at the level of the corpi striati and nucleus caudatus. Its disappearance at this level seems to play an essential part in the appearance of Parkinson's disease. *Noradrenalin* (or norepinephrin) is a hormone of the sympathetic nerve endings and is contained in the storage granules. *Adrenalin* (or epinephrin), together with noradrenalin, is particularly abundant in the medullary part of the suprarenal gland, which may be considered embryologically as a homologue of a nerve ganglion. It is the hormone of urgent response to aggression by the organism.

DNA and RNA

Deoxyribonucleic acid (DNA)

These molecules, which are the seat of genetic characters, are situated essentially in the cellular nucleus. Their structure is similar to RNA, with the difference that the sugar is a deoxyribose, and that the base uracil is replaced by thymine. They are found in the form of a double helical chain, with two complementary strands, one adenine nucleotide opposite a thymine nucleotide, a cytosine nucleotide opposite a guanine one. It seems that the biosynthesis of ribonucleic acids is also complementary to a fragment of the deoxyribonucleotic chain.

Ribonucleic acid (RNA)

A molecule present in all living cells, formed by the combination of a sugar (ribose) with 5 carbon atoms, of phosphoric acid and bases (adenine, guanine, uracil and cytosine). RNAs are located especially in particular regions of the intracellular membranes (endoplasmic reticulum) which make up the ribosomes. A small amount is free in the cytoplasm. Their main function is transporting amino acids to their specific site for protein synthesis (transfer RNA). Messenger RNAs are coded in the nucleus and appear to carry the message of the genome, inscribed on DNAs (deoxyribonucleic acids), to the enzymes which catalyse protein synthesis in the cytoplasm.

ELECTROENCEPHALIC WAVES

The electrical activity of the brain can be registered from the surface of the skull, which assumes global activity by its expression at the level of the cerebral cortex and in the various responses of the latter. It is also possible, by the insertion of electrodes in the different regions of the brain, after holes have been made in the skull, to collect the electrogenesis of these regions. This method is called stereotaxis.

H. Berger (1929) showed that according to the amplitude and rhythm

231

of the registered waves, the functioning of the region emitting them may be interpreted.

Full, slow waves are generally the expression of functional repose or an illness. On the other hand if they are less ample and the rhythm is rapid, strong neuronal activity is indicated. Alpha rhythm (the mechanics of which are still being debated) appears when the body is at rest. It has a frequency of between 6 and 15 cycles per second (on average, 10 cycles per second). It disappears when the eyes are opened.

Theta rhythm, which is quite characteristic of hippocampal activity, is of 4 to 7 cycles per second and of medium amplitude. Delta activity (1 to 3.5 cycles per second) is generally linked to a brain disorder. Epileptic fits provide recordings characterised by the presence of points of high voltage followed by waves (dart and dome).

During sleep, two phases are recognised: the predominant one of sleep, with slow full waves, and the other called paradoxical sleep (since on the electroencephalogram or EEG it is similar to the waking rhythm) which produces waves of weak amplitude and rapid rhythm.

During comas, the slowing down of the trace to a flat EEG, or even to nil, shows the brain's evolution towards death.

ENTROPY

The second law of thermodynamics (Carnot's principle) teaches us that the passage from one sort of energy to another proceeds in such a way that the total energy capable of producing work diminishes. This is because energy appears in two forms: kinetic energy, which is that of molecules animated by disordered movements, varying according to the temperature, and nil at absolute zero; and potential energy, which can be used to produce kinetic energy, and which appears in various forms, electric, calorific, chemical, light.

The second law tells us that there is a hierarchy of energy, ranged according to its capacity for being utilised to produce work; a system characterised by a high level of potential energy evolves in such a way that this level decreases to a lower value because of its transformation into kinetic energy, a degraded form of energy.

Since the work of Boltzmann, Maxwell and Gibbs this phenomenon can be expressed by saying that potential energy accompanies order and kinetic energy accompanies disorder, or in other words that potential energy carries more information than kinetic energy. Energy evolves towards the most probable state, and the second law of thermodynamics has become a statistical principle. It leads to the notion of entropy. The entropy of an isolated system is bound to go on increasing. Negentropy, the opposite of entropy, is then characterised by the appearance of increasing order.

ENVIRONMENT

All the objects and creatures immediately surrounding an individual. The individual is then considered to be situated in a "niche" formed by his environment. The phrase "internal medium" is also used, to describe all the physical, chemical and biological conditions in which an organism develops.

ENZYMES

Chemical reactions depend on conditions of temperature, pressure, pH etc., as do the output and even the possibility of the reaction. If these conditions were respected in life, then life would be impossible. Yet organisms are the site of a considerable number of chemical reactions, from which they draw the energy necessary to maintain their structure and their actions in the environment. Energy sources are taken from the external milieu (substrates), to be converted into "products" of a reaction. The reaction is catalysed by an enzyme, which is itself a product of the metabolism of the cell and is often highly specific for a given reaction. This reaction, when isolated from the chain of reactions it is part of (metalobic chain) is often a reaction of equilibrium.

The reaction is catalytic: the activation of two bodies reacting together is necessary to trigger it. Indeed, a reacting molecule starts at a set level of energy, passes through a higher level, and ends up at a level lower than the one from which it began. So the initial activation requires a certain quantity of energy (which is provided in general chemistry by, for example, heat). Because of its structure, the enzyme reduces the amount of activation energy to almost nothing. So it increases the speed of the reaction and make it possible at the temperature of ordinary organisms. Morover, it allows a greater number of molecules to be activated and the yield of the reaction is greater.

The enzyme is a product of the metabolism: it is a protein whose structure is engrammed in a particular gene which allows it to be synthesised when necessary.

The reaction is specific. This specificity can reach a high degree of subtlety. The enzyme may be specific to the substrate, to the type of chemical reaction, to the type of chemical bonding or even to the stereochemical configuration of the substrate.

The reaction is reversible. This is not a general characteristic. But the enzyme, while modifying the speed constant, does not modify the equilibrium constant of the reaction.

Michaelis and Menten (1913) suggested that the enzyme combines with the substrate, which in a first stage allows an intimate reaction: an exchange of electrons, for example. Then the products of the reaction appear and the enzyme, now released, is once more in its original state. They write of the enzymes for oxido-reduction:

"Supposing that the enzyme can combine not only with the substrate to be oxidised, but also with the oxidising agent. The specific structure of the enzyme leads to a definite orientation and a juxtaposition of the substances. If one molecule of each of these substances was in solution, the chance that an electron would be transferred from one to the other during a brief time of collusion would be practically nil. If these two substances are imprisoned in a ternary combination, the chances of a transfer are much greater."

Then again, stable molecules are usually surrounded by a ring of paired electrons (the electrons are generally linked in pairs). In this type of enzymatic reaction the electron exchange between the substance losing electrons and the one gaining them is carried out simultaneously by a *single* electron. The departure of the single electron causes the appearance of a "free radical", with an unpaired electron. This electron will try to pair by collecting an electron coming from an enzymatic reaction preceding it in the metabolic chain. Then, in its turn, it forms the product of the previous enzymatic reaction. Thus chains of enzymatic reactions are formed starting with a highly reduced substance and finishing with a highly oxidised substance. For glycolysis, the reduced substance will be glucose, which is rich in hydrogen molecules and therefore in electrons; the oxidised substance will be pyruvic acid, which will yield lactic acid on reduction (on gaining hydrogen and therefore electrons). For respiration, the substrate glucose will, after a long metabolic chain, discharge its electrons on to the oxygen molecule (O_2), a biradical lacking two electrons on its outer orbits.

HIPPOCAMPUS, HIPPOCAMPAL FORMATION OR ARCHICORTEX

This is a part of the old mammalian brain, and forms a ring surrounding the threshold of the hemisphere. The hippocampal fissure separates it from the neocortex. Related to the olfactory system in the lower vertebrates, in higher mammals and man it is just one of the essential elements relating the neocortex to the other most primitive cerebral formations, and it triggers complex motor and vegetative mechanisms. Its functioning dominates affectivity and the processes of memory.

HOMEOSTASIS

The tendency of the organism to keep its biological and physiological characteristics constant, particularly those of its internal medium.

HYPOTHALAMUS

Contains the essential centres of vegetative life. Their main function is to regulate the equilibrium of water (thirst), the thermic equilibrium, the metabolism of glucides and lipids, the arterial pressure, and the pigmentary and sleep functions. They also govern the hormonal function of the genital glands. The hypothalamus corresponds to the base of the brain on each side of the third ventricle and extends by the pituitary stem as far as the pituitary gland. In the hypothalamus an anterior region can be distinguished, containing three nuclei, and a posterior region also containing three nuclei. The anterior region is called trophotropic, the posterior one ergotropic (W. R. Hess, 1954), for the anterior region slows respiration and lowers arterial pressure while the posterior one increases blood pressure and has an accelerating effect. When stimulated it causes the suprarenal glands to release adrenalin.

It is the oldest and most primitive brain area, in which the inborn and most immediately indispensable forms of behaviour for survival are programmed, i.e. those which directly ensure the preservation of homeostasis.

MITOCHONDRIA

Intracellular organelles whose precise structure was revealed a few years ago, by different methods of isolation, fractioning and observation. They are the site of the main biochemical mechanisms which, in the animal cell, allow oxygen to be used as a receiver of electrons at the ends of biocatalytic chains. The arrangement of these reactions was supplied by H. Krebs, and the cycle he described (a tricarboxylic cycle or cycle of citric acid) is generally known as Krebs's cycle. The sequence of these reactions, starting with pyruvate, allows hydrogen molecules to be progressively removed from the substrates and ionised, i.e. the electron is separated from the proton.

The electron, charged with energy, then follows a molecular chain, known as chain of electron carriers, in which it gradually loses its energy in "small change" to "coenzymatic" molecules which allow the energy to be accumulated in the link of the third phosphorus of adenosine triphosphate (ATP). This molecule is the energy reserve which can be used immediately to preserve living structures and for their activity. Once it has lost its activation energy, the electron becomes attached to oxygen, taking the place of one of the missing electrons. When the two electrons are replaced, oxygen can then combine with the proton (hydrogen ion H⁺) to form a molecule of water (H₂O). But this final product is preceded by numerous intermediate free radical forms, whose oxidising role is undoubtedly of considerable biological importance, particularly in the ageing processes.

NEUROSES

The definition given by Ey, Bernard and Brisset (1967) is that of
— mental illnesses which are "minor" compared to psychoses,
— with preponderant subjective disturbances,
— and with a framework of more or less artificial and subconscious procedures against distress,
with full recognition of the numerous points which intervene between neuroses, which are mainly functional illnesses, and psychoses, illnesses which are considered mainly organic. Psychoanalysis has enriched the treatment of neuroses, for the "neurotic ego is defending itself against the internal danger of intrapsychic conflict". A distinction can be made between an undifferentiated or anxiety neurosis and differentiated neuroses: phobia, hysteria and obsessions.

PHOTON

A quantum of electromagnetic radiation. Radiation energy is a discontinuous series of minute (hypothetical) packets of energy, photons. Each of these packets when falling on matter is capable of expelling an electron, which allows the equivalence to be established between radiation energy and electrical energy, according to the expression $hf = \frac{1}{2} mv_2 = eV$, where h = Planck's constant; f = frequency of radiation; m = mass of the electron; e = its charge; v = its speed; V = potential difference.

PNEUMOGASTRIC OR VAGUS NERVE

Principal nerve of the parasympathetic system (10th cranial pair). The neurons of origin are situated in the oval cell body. The ganglion relay centres are usually situated in the effector organ itself, and not in the paravertebral region as for the sympathetic system. The pre- and postganglionic fibres release the same mediator at the nerve ending: acetylcholine. The vagus innervates most of the viscera — the heart (which it stops or slows down), the coronary vessels (which it constricts), the bronchi, the oesophagus, the stomach and intestines — except at the level of the sphincters, which it relaxes. It also has secretory functions in the stomach and pancreas and seems to be a vasodilator for certain narrow blood vessels. Its activity is in general opposed to that of the adrenergic sympathetic system.

PSYCHOSES

Mental illnesses characterised usually by a profound disintegration of the personality, with disturbances of perception, judgement and reasoning, of which the patient is unaware. Psychoses are usually long-lasting, but may include periods of lucidity. Psychotherapy has little effect. They include, among others, schizophrenia, dementia, and manic depressive psychosis.

SEROTONINE or 5-hydroxytryptamine (5-HT)

A hormone derived from tryptophane, found in large quantities in certain cells of the digestive canal and in brain tissue. Its effects are well-known, but its role and general significance are still obscure. It intervenes in the functioning of the brain. It is particularly plentiful in the limbic system, which leads one to suspect that it plays a role in affectivity and sleep. The diethylamide in lysergic acid (LSD) appears to be antagonistic to it. It is excreted in urine in the form of 5-hydroxy-indole-acetic acid.

SHAM RAGE

A behaviour in animals expressing a state of violent rage with no apparent cause in the environment. Is seen in the decorticated animal, whose lesions of the limbic system prevent the higher centres from controlling the activity of the hypothalamus.

SPLANCHNIC NERVES

The greater and lesser splanchnics arise from the sympathetic column of the spinal cord (lateral horn) between the 5th and 10th segment. Their fibres join and cross the paravertebral sympathetic ganglions without forming a relay centre, and join together in separate trunks. These are formed of pre-ganglionic fibres, which are therefore cholinergic, that is, they release acetylcholine at the nerve ending. They then join the semi-lunar (coeliac) ganglions. Certain fibres form a relay there, others join up directly with the suprarenal glands, whose medullary region they innervate. This medullary region secretes ardenalin and noradrenalin into the circulating blood. Thus this secretion is controlled by a cholinergic mechanism.

STRIATUM, NUCLEI STRIATI OR NUCLEUS CAUDATUS

A group of nuclei (nucleus caudatus, lenticular nucleus, which is itself divided into an external nucleus, the putamen, and two internal ones, the pallidum) surrounding the thalamus, from which they are separated by white fibres (of conduction): the internal capsule. They form a barrier between the cerebral cortex above and the other nervous structures below. This intermediary or paleocephalic brain can on its own play the role of a centre in relation to afferent and efferent fibres which are independent of the cortical system. Diagrammatically, it is part of a primitive brain in which the afferences to the thalamus can be projected to the centres of origin of the efferent pathways (nuclei striati) by means of thalamostriated pathways. The whole forms an autonomous reflex group, anatomically and physiologically individualised. Its relative dependence with regard to the neocephalus is phylogenetically a secondary process. The nuclei striati thus form a primitive motor brain. One tends, on the basis of their anatomical connections and their function and embryological origin, to divide the nuclei striati into two masses: one, the pallidum or paleostriatum, includes the two internal nuclei of the lenticular; the other, the neostriatum, includes the nucleus caudatus and the external lenticular nucleus, also called the putamen. At present the paleostriatum is called the pallidum and the neostriatum is called the striatum.

SUPRARENALS

Endocrine glands, situated on the upper pole of the kidneys. Two regions can be distinguished: the medullary, which secrets noradrenalin and especially adrenalin (norepinephrin and epinephrin), and the cortical, which secretes the cortico-suprarenal hormones. These hormones act in some cases mainly on the cell metabolism of glucose (glucocorticoids), whose type is cortisol (hydrocortisone); others on hydro-mineral balance (concentration in tissues and humours of mineral salts and water) or mineralocorticoids, whose type is desoxycorticosterone (DOC). The metabolism of the cortico-suprarenal cells leading to the secretion of the suprarenal corticoids is dependent on the secretion of a hormone by the ante-hypophysis (ACTH), which is itself triggered by the hypothalamus.

The medullary secretion of adrenalin is directly dependent on stimulation of the splanchnics.

The most recent region of the brain, which appeared in the early mammals (hedgehog) and which is essential for memory, itself indispensable for affectivity and training. The limbic system is thus behind the acquisition of automatisms, both of action and thought.

THALAMUS

This sensory centre of the paleocephalus is a central nerve mass which is subcortical, voluminous, and formed of phylogenetically distinct nuclei:

"The paleothalamus, an old sensory nucleus, as its name indicates, is situated in the extension of the brain stem which provides it with most of its afferences. It maintains paleoencephalic connections with the nuclei striati.

"The neothalamus lies above the paleothalamus, receives its afferences from it and sends its afferent fibres towards the neochepalus, that is towards the cerebral cortex.

"Furthermore, paleo- and neothalamus are enveloped and crossed by diffuse grey formations, or thalamic reticuli, of an identical nature to those already studied in the brain stem. Certain authors consider them to be the remains of the primitive partition of the diencephalon of fish or reptiles. The reticulated thalamic formation would thus represent an archithalamus. An attractive idea, suggesting the existence within the thalamus of structures which are not only hierarchised but also very specialised: the archithalamus as a diffuse activator, the paleothalamus as a relay centre for the sensory tracts and suprasegmentary reflexes, the neothalamus as linked to the cerebral cortex, which is the centre for the sorting and regrouping of messages from the levels below. Yet the idea of an archithalamus, i.e. a primitive thalamus, is still disputed (Kühlenbeck) for the very reason that the reticulated thalamic formation is considerably developed in man." (Delmas, A., *Voies et centres nerveux,* Paris 1969.)

THERMODYNAMICS

Branch of physics dealing with relationships between thermal phenomena and mechanical phenomena (work or mechanical energy).

Some other titles in the MOTIVE series

Michel Raptis
Revolution and Counter-Revolution in Chile

Henri Lefebvre
The Survival of Capitalism

Franz Jakubowski
Ideology and Superstructure in Historical Materialism

Agnes Heller
The Theory of Need in Marx

Jiri Pelikan
Socialist Opposition in Eastern Europe

Mihaly Vajda
Fascism as a Mass Movement

Hilda Scott
Women and Socialism — Experiences from Eastern Europe

Andras Hegedus
Socialism and Bureaucracy

Andras Hegedus, Agnes Heller, Maria Markus, Mihaly Vajda
The Humanisation of Socialism

Bill Lomax
Hungary 1956

Simon Leys
The Chairman's New Clothes — Mao and the Cultural Revolution

Guy Hocquenghem
Homosexual Desire